# Creo™ Parametric 2.0

## Louis Gary Lamit
With technical assistance from **James Gee**

De Anza College

CENGAGE Learning®

Australia • Brazil • Japan • Korea • Mexico • Singapore • Spain • United Kingdom • United States

*Creo™ Parametric 2.0*

**Louis Gary Lamit**

Publisher: Timothy L. Anderson

Senior Developmental Editor: Hilda Gowans

Editorial Assistant: Tanya Altieri

Marketing Manager: Lauren Betsos

Production Management, and
Composition: PreMediaGlobal

Media Editor: Chris Valentine

Rights Acquisition Director: Audrey
Pettengill

Rights Acquisition Specialist, Text and
Images: Amber Hosea

Senior Manufacturing Planner: Doug Wilke

Senior Art Director: Michelle Kunkler

Cover Designer: c miller design

Cover Images: © Mikhail/Shutterstock;
© Vitalii Culibacinii/iStockphoto;
© penfold/iStockphoto

For product information and technology assistance, contact us at
**Cengage Learning Customer & Sales Support, 1-800-354-9706**

For permission to use   material from this text or product, submit all requests online at **www.cengage.com/permissions**
Further permissions questions can be emailed to
**permissionrequest@cengage.com**

Library of Congress Control Number: 2012952055

ISBN-13: 978-1-285-19071-6
ISBN-10: 1-285-19071-8

**Cengage Learning**
200 First Stamford Place, Suite 400
Stamford, CT 06902
USA

Cengage Learning is a leading provider of customized learning solutions with office locations around the globe, including Singapore, the United Kingdom, Australia, Mexico, Brazil, and Japan. Locate your local office at: **www.cengage.com/region**

Cengage Learning products are represented in Canada by Nelson Education, Ltd.

For your course and learning solutions, visit
**www.cengage.com/engineering**

Purchase any of our products at your local college store or at our preferred online store **www.cengagebrain.com**

Printed in the United States of America
1 2 3 4 5 6 7  16 15 14 13 12

# Table of Contents

# Downloadable Lessons

# About the Author and technical assistant

*Louis Gary Lamit* is currently a full time instructor and CDI (CAD) department chair at De Anza College (since 1984) in Cupertino, CA, where he teaches Creo Parametric 2.0. He is the founder of Scholarships for Veterans at www.scholarshipsforveterans.org. Mr. Lamit has worked as a drafter, designer, numerical control (NC) programmer, technical illustrator, and engineer in the automotive, aircraft, and piping industries. A majority of his work experience is in the area of mechanical and piping design. He started as a drafter in Detroit (as a job shopper) in the automobile industry, doing tooling, dies, jigs and fixture layout, and detailing at Koltanbar Engineering, Tool Engineering, Time Engineering, and Premier Engineering for Chrysler, Ford, AMC, and Fisher Body. Mr. Lamit has worked at Remington Arms and Pratt & Whitney Aircraft as a designer, and at Boeing Aircraft and Kollmorgan Optics as an NC programmer and aircraft engineer. He also owns and operates his own consulting firm (CAD-Resources.com- Lamit and Associates), and has been involved with advertising, and patent illustration. He is the author of over 40 books, journals, textbooks, workbooks, tutorials, and handbooks, including children's journals and books (www.walkingfishbooks.com). Mr. Lamit received a BS degree from Western Michigan University in 1970 and did Masters' work at Wayne State University and Michigan State University. He has also done graduate work at the University of California at Berkeley and holds an NC programming certificate from Boeing Aircraft. Since leaving industry, Mr. Lamit has taught at all levels (Melby Junior High School, Warren, Mi.; Carroll County Vocational Technical School, Carrollton, Ga.; Heald Engineering College, San Francisco, Ca.; Cogswell Polytechnical College, San Francisco and Cupertino, Ca.; Mission College, Santa Clara, Ca.; Santa Rosa Junior College, Santa Rosa, Ca.; Northern Kentucky University, Highland Heights, Ky.; and De Anza College, Cupertino, Ca.). His textbooks include:

- *Industrial Model Building*, with Engineering Model Associates, Inc. (1981),
- *Piping Drafting and Design* (1981),
- *Piping Drafting and Design Workbook* (1981),
- *Descriptive Geometry* (1983),
- *Descriptive Geometry Workbook* (1983), and
- *Pipe Fitting and Piping Handbook* (1984), Prentice-Hall.
- *Drafting for Electronics* (3rd edition, 1998),
- *Drafting for Electronics Workbook* (2nd edition 1992), and
- *CADD* (1987), Charles Merrill (Macmillan-Prentice-Hall Publishing).
- *Technical Drawing and Design* (1994),
- *Technical Drawing and Design Worksheets and Problem Sheets* (1994),
- *Principles of Engineering Drawing* (1994),
- *Fundamentals of Engineering Graphics and Design* (1997),
- *Engineering Graphics and Design with Graphical Analysis* (1997), and
- *Engineering Graphics and Design Worksheets and Problem Sheets* (1997), West Publishing (ITP/Delmar).
- *Basic Pro/ENGINEER in 20 Lessons* (1998) (Revision 18) and
- *Basic Pro/ENGINEER (with references to PT/Modeler)* (1999), PWS.
- *Pro/ENGINEER 2000i* (1999), and
- *Pro/ENGINEER 2000i$^2$ (Pro/NC and Pro/SHEETMETAL)* (2000), Brooks/Cole Publishing (ITP).
- *Pro/ENGINEER Wildfire* (2003), Brooks/Cole Publishing (ITP).
- *Introduction to Pro/ENGINEER Wildfire 2.0* (2004), SDC.
- *Moving from 2D to 3D CAD for Engineering Design* (2007), BookSurge, eBook by MobiPocket.
- *Pro/ENGINEER Wildfire 3.0 Tutorial* (2007), BookSurge, eBook by MobiPocket.
- *Pro/ENGINEER Wildfire 3.0* (2007), Cengage.
- *Pro/ENGINEER Wildfire 4.0 Tutorial* (2008), BookSurge eBook by MobiPocket.
- *Pro/ENGINEER Wildfire 4.0* (2008), Cengage.
- *Pro/ENGINEER Wildfire 5.0* (2010), Cengage.
- *Creo Parametric (2012), Cengage*
- *Creo Parametric 2.0 (2014), Cengage*

*James Gee* is currently a part time instructor at De Anza College, where he teaches Creo Parametric 2.0 (application modules including Sheetmetal, Cabling, Surfaces, and Mechanica) and NX. Mr. Gee graduated from the University of Nevada- Reno with a BSME. He has worked in the Aerospace industry for Lockheed Missiles and Space Company, Sunnyvale, Ca.; Space Systems/Loral, Palo Alto, Ca.; and BAE Systems in San Jose, Ca. Mr. Gee has assisted in checking and editing the Pro/ENGINEER/Creo series of textbooks by Louis Gary Lamit.

# Preface

**Creo Parametric 2.0** is one of the most widely used CAD/CAM software programs in the world today. Any aspiring engineer will greatly benefit from the knowledge contained herein, while in school or upon graduation as a newly employed engineer.

   *This is the second version of Creo Parametric by PTC*. Significant changes, upgrades, and new capabilities including have made this software a unique product. *This is not a revised textbook* but a new book covering all the necessary subjects needed to master this high-level CAD software. There are few if any comprehensive texts on this subject so we hope this text will fill the needs of both schools and professionals alike.

   The text involves creating a new part, an assembly, or a drawing, using a set of commands that walk you through the process systematically. Lessons and Projects all come from industry and have been tested for accuracy and correctness as per engineering standards. Projects are downloadable as a PDF with live links and 3D embedded models.

**Lessons 13-18** (downloadable PDF)**, Video Lessons 19-22,** and **Lesson Projects** (downloadable PDF) are not included in the printed version to keep the length and cost to the user down- lessons and projects not in the printed portion can be downloaded at:
**http://www.cad-resources.com/Creo_Parametric_Cengage.html?1341766955310**

A **complete set of Recorded Lectures for this book is available for free at The Cengage Author Resource page.** The Lectures are in WMV format and run between 15 and 40 minutes for each lesson/project. The Lectures are recorded using a commercial version of Creo Parametric 2.0 and are the exact content presented in the classroom by the author.
**http://www.cad-resources.com/Creo_Parametric_Cengage.html?1341766955310**

**Creo Parametric 2.0 Free Schools Edition software is available at www.ptc.com.**

## Contact

If you wish to contact the author concerning orders, questions, changes, additions, suggestions, comments, or to get on our email list, please send an email to one of the following:

Web Site: **www.cad-resources.com**
Email: **lgl@cad-resources.com**

## Acknowledgments

**Max Gilliland**- instructional associate for CAD at De Anza College has been essential to the CAD program. Besides assisting in the classroom, he maintains the software and hardware for the CAD program and for my publications.

**Alistair Fraser**- at PTC for continuing support of our program at De Anza College and for my publications of PTC products.

# On campus and Online CAD Classes at De Anza College

All CAD classes at **De Anza College** are available on campus and **on line** at **www.deanza.edu/cdi/**

**Creo Parametric 2.0**:
- CDI70 Beginning
- CDI71 Intermediate
- CDI72 Advanced
- CDI73 Sheetmetal
- CDI74 Surface

**SolidWorks**:
- CDI60 Beginning
- CDI61 Intermediate
- CDI62 Advanced
- CDI63 Surface

**AutoDESK**:
- **AutoCAD**: CDI80 Beginning, CDI81 Intermediate
- **Inventor**: CDI85
- **Revit**: CDI83

**NX**: CDI58

**CATIA**: CDI95

**GD&T:** CDI51

## Dedication

This book is dedicated to my first grandson: **James**

*gate gate pāragate pārasagate bodhi svāhā*

## Donations and Scholarships for Veterans (SFV)

A portion of this text's profits go to my scholarship fund at Foothill-De Anza Community College District (FHDA Foundation). Your contributions provide extra scholarships as funds are available.

SFV provides funding for tuition and fees. Scholarships are available to any qualified veteran of the Army, Navy, Air Force, Marines, or Coast Guard. Scholarships are administered by the local college foundations. No administration fees are taken by Scholarships for Veterans.

All costs associated with Scholarships for Veterans are borne by Lamit and Associates and CAD-Resources.com. **www.cad-resources.com/SFV.html**

# Introduction

This text guides you through parametric design using Creo Parametric 2.0™. While using this text, you will create individual parts, assemblies, and drawings.

**Parametric** can be defined as *any set of physical properties whose values determine the characteristics or behavior of an object*. **Parametric design** enables you to generate a variety of information about your design: its mass properties, a drawing, or a base model. To get this information, you must first model your part design.

Parametric modeling philosophies used in Creo Parametric 2.0 include the following (Fig. 1):

*Feature-Based Modeling* Parametric design represents solid models as combinations of engineering features.

*Creation of Assemblies* Just as features are combined into parts, parts may be combined into assemblies.

*Capturing Design Intent* The ability to incorporate engineering knowledge successfully into the solid model is an essential aspect of parametric modeling.

**Figure 1** Parts and Assembly Design

## Parametric Design

Parametric design models are not drawn so much as they are *sculpted* from solid volumes of materials. To begin the design process, analyze your design. Before any work is started, take the time to *tap* into your own knowledge bank and others that are available. **T**hink, **A**nalyze, and **P**lan. These three steps are essential to any well-formulated engineering design process.

Break down your overall design into its basic components, building blocks, or primary features. Identify the most fundamental feature of the object to sketch as the first, or base, feature. Varieties of **base features** can be modeled using extrude, revolve, sweep, and blend tools.

**Sketched features** (*extrusions, sweeps, etc.*) and pick-and-place features called **referenced features** (*holes, rounds, chamfers, etc.*) are normally required to complete the design. With the SKETCHER, you use familiar 2D entities (points, lines, rectangles, circles, arcs, splines, and conics) (Fig. 2). There is no need to be concerned with the accuracy of the sketch. Lines can be at differing angles, arcs and circles can have unequal radii, and features can be sketched with no regard for the actual objects' dimensions. In fact, exaggerating the difference between entities that are similar but not exactly the same is actually a far better practice when using the SKETCHER.

**Figure 2** Sketching

**Geometry assumptions and constraints** will close ends of connected lines, align parallel lines, and snap sketched lines to horizontal and vertical (orthogonal) orientations. Additional constraints are added by means of **parametric dimensions** to control the size and shape of the sketch.

**Features** are the basic building blocks you use to create an object (Fig. 3). Features "understand" their fit and function as though "smarts" were built into the features themselves. For example, a hole or cut feature "knows" its shape and location and the fact that it has a negative volume. As you modify a feature, the entire object automatically updates after regeneration. The idea behind feature-based modeling is that the designer constructs an object so that it is composed of individual features that describe the way the geometry is supposed to behave if its dimensions change. This happens quite often in industry, as in the case of a design change. Feature-based modeling is diagramed in Figure 4.

**Figure 3** Feature Design

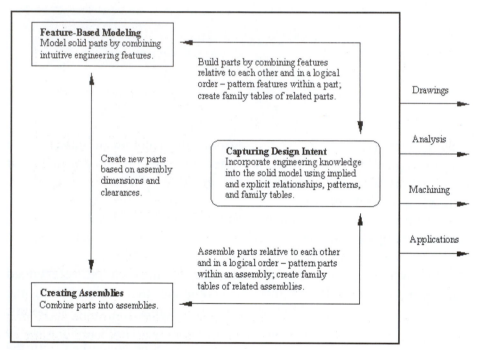

**Figure 4** Feature-based Modeling

**Parametric modeling** is the term used to describe the capturing of design operations as they take place, as well as future modifications and editing of the design. The order of the design operations is significant. Suppose a designer specifies that two surfaces be parallel, such that surface two is parallel to surface one. Therefore, if surface one moves, surface two moves along with surface one to maintain the specified design relationship. The surface two is a **child** of surface one in this example. Parametric modeling software allows the designer to **reorder** the steps in the object's creation.

Various types of features are used as building blocks in the progressive creation of solid objects. Figures 5(a-c) illustrates base features, datum features, sketched features, and referenced features. The "chunks" of solid material from which parametric design models are constructed are called **features**.

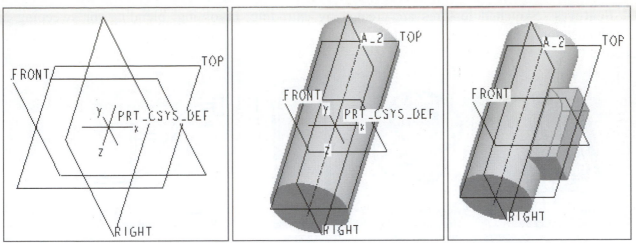

**Figures 5(a-c)** Features

Features generally fall into one of the following categories:

**Base Feature** The base feature is normally a set of datum planes referencing the default coordinate system. The base feature is important because all future model geometry will reference this feature directly or indirectly; it becomes the root feature. Changes to the base feature will affect the geometry of the entire model [Figs. 6(a-c)].

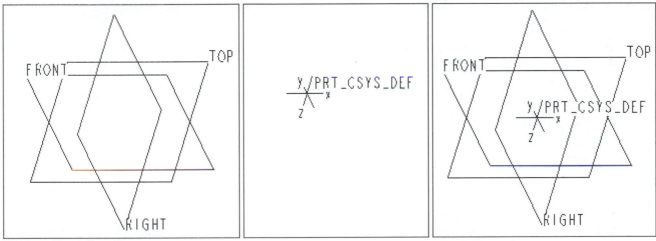

**Figures 6(a-c)** Base Features

**Datum Features** Datum features (lines, axes, curves, and points) are generally used to provide sketching planes and contour references for sketched and referenced features. Datum features do not have volume or mass and may be visually hidden without affecting solid geometry (Fig. 7).

**Figure 7** Datum Curves

**Sketched Features** Sketched features are created by extruding, revolving, blending, or sweeping a sketched cross section. Material may be added or removed by protruding or cutting the feature from the existing model (Fig. 8).

**Figure 8** Sketched Features

**Referenced Features** Referenced features (rounds, holes, shells, and so on) utilize existing geometry for positioning and employ an inherent form; they do not need to be sketched [Figs. 9(a-b)].

**Figures 9(a-b)** Referenced Features- Shell and Round *(Spout is a Swept Blend Feature)*

A wide variety of features are available. These tools enable the designer to make far fewer changes by capturing the engineer's design intent early in the development stage (Fig. 10).

**Figure 10** Parametric Designed Part

# Fundamentals

The design of parts and assemblies, and the creation of related drawings, forms the foundation of engineering graphics. When designing with Creo Parametric 2.0, many of the previous steps in the design process have been eliminated, streamlined, altered, refined, or expanded. The model you create as a part forms the basis for all engineering and design functions.

The part model contains the geometric data describing the part's features, but it also includes non-graphical information embedded in the design itself. The part, its associated assembly, and the graphical documentation (drawings) are parametric. The physical properties described in the part drive (determine) the characteristics and behavior of the assembly and drawing. Any data established in the assembly mode, in turn, determines that aspect of the part and, subsequently, the drawings of the part and the assembly. In other words, all the information contained in the part, the assembly, and the drawing is interrelated, interconnected, and parametric (Fig. 11).

**Figure 11** Assembly Drawing, Part, and Assembly

## Part Design

In many cases, the part will be the first component of this interconnected process. The *part* function in Creo Parametric 2.0 is used to design components.

During part design (Fig. 12), you can accomplish the following:

- Define the base feature
- Define and redefine construction features to the base feature
- Modify the dimensional values of part features (Fig. 13)
- Embed design intent into the model using tolerance specifications and dimensioning schemes
- Create pictorial and shaded views of the component
- Create part families (family tables)
- Perform mass properties analysis and clearance checks
- List part, feature, layer, and other model information
- Measure and calculate model features
- Create detail drawings of the part

5

**Figure 12** Part Design

**Figure 13** Pick on the **.56** dimension and type in a new value

## Establishing Features

The design of any part requires that the part be *confined*, *restricted*, *constrained*, and *referenced*. In parametric design, the easiest method to establish and control the geometry of your part design is to use three datum planes. Creo Parametric 2.0 automatically creates the three **primary datum planes**. The default datum planes (**RIGHT**, **TOP**, and **FRONT**) constrain your design in all three directions.

  **Datum planes** are infinite planes located in 3D model mode and are associated with the object that was active at the time of their creation. To select a datum plane, you can pick on its name or anywhere on the perimeter edge. Datum planes are *parametric*--geometrically associated with the part. Parametric datum planes are associated with and dependent on the edges, surfaces, vertices, and axes of a part. Datum planes are used to create a reference on a part that does not already exist. For example, you can sketch or place features on a datum plane when there is no appropriate planar surface. You can also dimension to a datum plane as though it were an edge. In Figure 14, three **default datum planes** and a **default coordinate system**s were created when a new part (**PRT0001.PRT**) was started using the default template. Note that in the **Model Tree** window, they are the first four features of the part, which means that they will be the *parents* of the features that follow.

**Figure 14** Default Datum Planes and Coordinate System

## Datum Features

**Datum features** are planes, axes, and points you use to place geometric features on the active part. Datums other than defaults can be created at any time during the design process.

As we have discussed, there are three (primary) types of datum features: **datum planes** (Fig. 14), **datum axes**, and **datum points** (there are also *datum curves* and *datum coordinate systems*). You can display all types of datum features, but they do not define the surfaces or edges of the part or add to its mass properties. In Figure 15, a variety of datum planes are used in the creation of the casting.

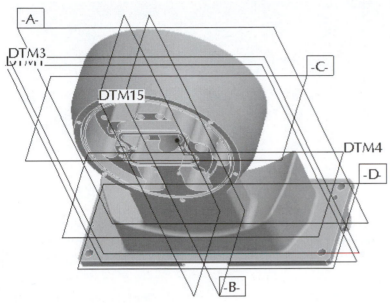

**Figure 15** Datums in Part Design

Specifying constraints that locate it with respect to existing geometry creates a datum. For example, a datum plane might be made to pass through the axis of a hole and parallel to a planar surface. Chosen constraints must locate the datum plane relative to the model without ambiguity. You can also use and create datums in assembly mode.

Besides datum planes, datum axes and datum points can be created to assist in the design process. You can also automatically create datum axes through cylindrical features such as holes and solid round features by setting this as a default in your Creo Parametric 2.0 configuration file. The part in Figure 16 shows the default axes for a variety of holes.

**Figure 16** Feature Default Datum Axes

## Parent-Child Relationships

Because solid modeling is a cumulative process, certain features must, by necessity precede others. Those that follow must rely on previously defined features for dimensional and geometric references. The relationships between features and those that reference them are termed *parent-child relationships*. Because children reference parents, parent features can exist without children, but children cannot exist without their parents. This type of CAD modeler is called a history-based system. Using Creo Parametric 2.0's information command will list a model's parent-child references and dependencies (Fig. 17).

**Figure 17** Model Information

**Figure 18** Parent-Child Information

The parent-child relationship (Fig. 18) is one of the most powerful aspects of parametric design. When a parent feature is modified, its children are automatically recreated to reflect the changes in the parent feature's geometry. It is essential to reference feature dimensions so that design modifications are correctly propagated through the model/part. Any modification to the part is automatically propagated throughout the model (Fig. 19) and will affect all children of the modified feature.

**Figure 19** Original Design and Modification

# Capturing Design Intent

A valuable characteristic of any design tool is its ability to *render* the design and at the same time capture its *intent* (Fig. 20). Parametric methods depend on the sequence of operations used to construct the design. The software maintains a *history of changes* the designer makes to specific parameters. The point of capturing this history is to keep track of operations that depend on each other. Whenever Creo Parametric 2.0 is told to change a specific dimension, it can update all operations that are referenced to that dimension.

For example, a circle representing a bolt hole circle may be constructed so that it is always concentric to a circular slot. If the slot moves, so does the bolt circle. Parameters are usually displayed in terms of dimensions or labels and serve as the mechanism by which geometry is changed. The designer can change parameters manually by changing a dimension or can reference them to a variable in an equation (**relation**) that is solved either by the modeling program itself or by external programs such as spreadsheets.

Features can also store non-graphical information. This information can be used in activities such as drafting, numerical control (NC), finite-element analysis (FEA), and kinematics analysis.

Capturing design intent is based on incorporating engineering knowledge into a model by establishing and preserving certain geometric relationships. The wall thickness of a pressure vessel, for example, should be proportional to its surface area and should remain so, even as its size changes.

**Figure 20** Capturing Design Intent

Parametric designs capture relationships in several ways:

**Implicit Relationships**   Implicit relationships occur when new model geometry is sketched and dimensioned relative to existing features and parts. An implicit relationship is established, for instance, when the section sketch of a tire (Fig. 21) uses rim edges as a reference.

**Figure 21** Tire and Rim

**Patterns**   Design features often follow a geometrically predictable pattern. Features and parts are patterned in parametric design by referencing either construction dimensions or existing patterns. One example of patterning is a wheel hub with spokes (Fig. 22). First, the spoke holes are radially patterned. The spokes can then be strung by referencing this pattern.

**Figure 22** Patterns

Modification to a pattern member affects all members of that pattern. This helps capture design intent by preserving the duplicate geometry of pattern members.

The modeling task is to incorporate the features and parts of a complex design while properly capturing design intent to provide flexibility in modification. Parametric design modeling is a synthesis of physical and intellectual design (Fig. 23).

**Figure 23** Relations

**Explicit Relations** Whereas implicit relationships are implied by the feature creation method, the user mathematically enters an explicit relation. This equation is used to relate feature and part dimensions in the desired manner. An explicit relation (Fig. 24) might be used, for example, to control sizes on a model.

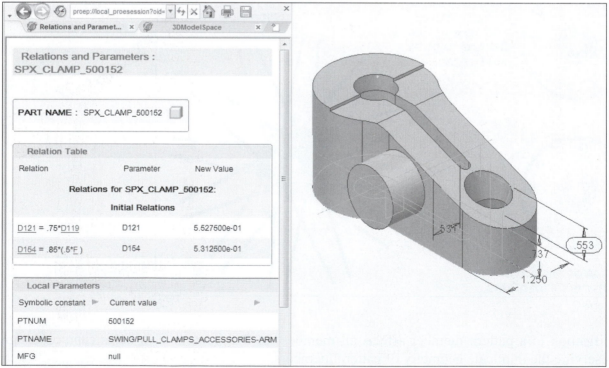

**Figure 24** Adding Relations

**Family Tables** Family tables are used to create part families [Figs. 25(a-c)] from generic models by tabulating dimensions or the presence of certain features or parts. A family table might be used, for example, to catalog a series of couplings with varying width and diameter as shown in Figure 26.

Figures 25(a-c) Family of Parts- Coupling-Fitting

| Type | Instance Name | Common Name | d30 | d46 | F1068 [CUT] | F829 [SLOT] | F872 [SLOT] | F2620 COPIED_G... |
|------|---------------|-------------|------|--------|------|------|------|------|
| | COUPLING-FITTI... | | 2.06 | 0.3130 | Y | Y | Y | Y |
| | CPLA | | 3.00 | 0.5000 | N | N | N | N |
| | CPLB | | 3.25 | 0.6250 | N | Y | N | N |
| | CPLC | | 3.50 | 0.7500 | Y | N | Y | Y |

Figure 26 Family Table for Coupling-Fitting

## Assemblies

Just as parts are created from related features, **assemblies** are created from related parts. The progressive combination of subassemblies, parts, and features into an assembly creates parent-child relationships based on the references used to assemble each component (Fig. 27).

The *Assembly* functionality is used to assemble existing parts and subassemblies.

Figure 27 Clamp Assembly and Exploded Clamp Assembly

During assembly creation, you can:

- Simplify a view of a large assembly by creating a simplified representation
- Perform automatic or manual placement of component parts
- Create an exploded view of the component parts
- Perform analysis, such as mass properties and clearance checks
- Modify the dimensional values of component parts
- Define assembly relations between component parts
- Create assembly features
- Perform automatic interchange of component parts
- Create parts in Assembly mode
- Create documentation drawings of the assembly

Just as features can reference part geometry, parametric design also permits the creation of parts referencing assembly geometry. **Assembly mode** allows the designer both to fit parts together and to design parts based on how they should fit together. In Figure 28, an assembly *Bill of Materials* report is generated.

**Figure 28** BOM Report

# Drawings

You can create drawings of all parametric design models (Fig. 29). All model views in the drawing are *associative:* if you change a dimensional value in one view, other drawing views update accordingly. Moreover, drawings are associated with their parent models. Any dimensional changes made to a drawing are automatically reflected in the model. Any changes made to the model (e.g., addition of features, deletion of features, dimensional changes, and so on) in Part, Sheet Metal, Assembly, or Manufacturing modes are also automatically reflected in their corresponding drawings.

**Figure 29** Ballooned Exploded View Assembly Drawing with Bill of Materials (BOM)

The **Drawing** functionality is used to create annotated drawings of parts and assemblies. During drawing creation, you can:

- Add views of the part or assembly
- Show existing dimensions
- Incorporate additional driven or reference dimensions
- Create notes on the drawing
- Display views of additional parts or assemblies
- Add sheets to the drawing
- Create draft entities on the drawing
- Balloon components on an assembly drawing (Fig. 30)
- Create an associative BOM

You can annotate the drawing with notes, manipulate the dimensions, and use layers to manage the display of different items on the drawing. The module **Pro/DETAIL** can be used to extend the drawing capability or as a stand-alone module allowing you to create, view, and annotate models and drawings. Pro/DETAIL supports additional view types and multi-sheets and offers commands for manipulating items in the drawing and for adding and modifying different kinds of textural and symbolic information. In addition, the abilities to customize engineering drawings with sketched geometry, create custom drawing formats, and make numerous cosmetic changes to the drawing are available.

**Figure 30** Ballooned Assembly Drawing

**Drawing mode** in parametric design provides you with the basic ability to document solid models in drawings that share a two-way associativity.

Changes that are made to the model in Part mode or Assembly mode will cause the drawing to update automatically and reflect the changes. Any changes made to the model in Drawing mode will be immediately visible on the model in Part and Assembly modes. The model shown in Figure 31 has been detailed in Figure 32.

**Figure 31** Angle Frame Model

**Figure 32** Angle Frame Drawing

16

# Flexible Modeling

Flexible modeling allows you to experiment with your design without being committed to changes. You can make explicit modifications to selected geometry while ignoring pre-existing relationships. With this new capability within Creo Parametric 2.0 you can utilize some basic Creo Direct 2.0 capabilities. The ribbon in Figure 33 shows the new interface. Design changes are done directly on the model without parent-child relationships being imbedded (Fig. 34).

**Figure 33** Flexible Modeling Ribbon

**Figure 34** Design experimentation using Flexible Modeling

## Using the Text

The text utilizes a variety of command boxes and descriptions to lead you through construction sequences. Also, see the downloadable Creo Parametric 2.0 Quick Reference Cards at www.cad-resources.com. The following icons, symbols, shortcut keys, and conventions will be used *(command sequences are always in a box)*:

Click: **View** tab > **Display Style ▾** > **Shading With Edges** > **Standard Orientation** > **Model** tab > **Revolve** > in the Graphics Window, press **RMB** > ☑ **Remove Material** > press **RMB** > **Define Internal Sketch** > > Sketch Plane--- Plane: select datum **FRONT** from the Model Tree **FRONT** > Reference: select datum **TOP** > Orientation: **TOP** > **Sketch**

## *Commands:*

- **>**  Continue with command sequence or screen picks using **LMB**

- ⌃ Line Chain  Icon (with description) indicates command to pick using **LMB**

## *Mouse or keyboard terms used in this text:*

- **LMB**  Left Mouse Button
  - or "**Pick**"  term used to direct an action (i.e., "Pick the surface")
  - or "**Click**"  term used to direct an action (i.e., "Click on the icon")
  - or "**Select**"  term used to direct an action (i.e., "Select the feature")

- **MMB**  Middle Mouse Button (accept the current selection or value)
  - or **Enter**  press **Enter** key to accept entry

  - or ✓  Click on this icon to accept entry

- **RMB**  Right Mouse Button
  *Click* (toggles to next selection)
  *Press and hold* (provides a list of available commands)

## Shortcut Keys:

| Keystroke | Action |
|---|---|
| **ALT** | •Disables filters temporarily during selection when the pointer is in the graphics window. Hold down the ALT key during selection.<br><br>•Activates KeyTips on the Ribbon. With your pointer anywhere in the Creo window (Ribbon, Graphics Window, Model Tree, etc.), press the ALT key. |
| **CTRL** | Temporarily hides the Live Toolbar in 3D part and assembly modes. |
| **CTRL+A** | Activates a window. |
| **CTRL+C** | Copies the selection to the clipboard. |
| **CTRL+D** | Displays a model in standard view. |
| **CTRL+N** | Opens the **New** dialog box. |
| **CTRL+O** | Opens the **File Open** dialog box. |
| **CTRL+R** | Repaints a window. |
| **CTRL+S** | Opens the **Save Object** dialog box. |
| **CTRL+V** | Pastes the clipboard contents to the selected location. |
| **CTRL+Y** | Performs a single step redo operation. |
| **CTRL+Z** | Performs a single step undo operation. |
| **ESC** | Performs one of the following operations depending on the modeling context:<br>• Cancels an operation<br>• Clears the selection in the graphics window<br>• Cancels a dragger operation<br>• Closes the Live Toolbar<br>• Cancels the currently active tool |
| **ENTER** | Accepts the changes and closes the Live Toolbar in 3D and assembly modes. Also clears the selection in the graphics window. |
| **F1** | Opens **Help** for the current context. |
| **G** | Hides the current guide for the geometry in 2D mode. |
| **L** | Locks or unlocks the current guide in 2D mode. |
| **P** | Shows or hides precision panels in 2D mode temporarily. |
| **S** | Switches between the Line and Arc Creation modes in 2D mode. |
| **TAB** | Pre-highlights a set of surfaces for selection based on shape selection rules. Refer to the Creo Parametric 2.0 Flexible Modeling Help for more information on shape selection rules. |

# Text Organization

- **Lesson 1** has you create two simple parts, an assembly, and two drawings using default settings.
- **Lesson 2** introduces Creo Parametric 2.0's interface and embedded Browser.
- **Lesson 3** provides uncomplicated instructions to model a variety of simple-shaped parts.
- **Lessons 4** and **5** involve part modeling using a variety of commands and tools.
- **Lessons 6** through **12** involve modeling the parts, creating the assembly [Fig. 35(a)], and documenting a design with detail and assembly drawings [Fig. 35(b)].
- **Lessons 13-18, Video Lessons 19-22 (Download)** Instructions to model and document parts using more advanced features and techniques. Available as a download to minimize the size and cost of the text.
- **Projects (Download)** Provides over 100 pages of project drawings as a PDF download so as to minimize the size and cost of the text.

**Figure 35(a)** Clamp Assembly

**Figure 35(b)** Clamp Assembly Drawing

# Lesson 1 Creo Parametric 2.0 Overview

**Figure 1.1** Pin Model, Plate Model, Assembly Model, Plate Drawing, and Assembly Drawing

## OBJECTIVES

* Model two parts
* Assemble parts to create an assembly
* Construct a part drawing
* Make an assembly drawing

## REFERENCES AND RESOURCES

For **Resources** go to **www.cad-resources.com** > click on the image of your Book cover

* Lesson 1 Lecture
* Lesson 1 3D models embedded in a PDF
* Creo Parametric 2.0 Quick Reference Card
* Creo Parametric 2.0 Configuration Options

## Creo Parametric 2.0 Overview

This lesson (Fig. 1.1) will allow you to experience the part (Fig. 1.2), assembly, and drawing modes of Creo Parametric 2.0 by creating two simple parts, assembling them, and creating drawings. A minimum of explanation and instruction is provided here. This Creo Parametric 2.0 lesson will quickly get you up and running.

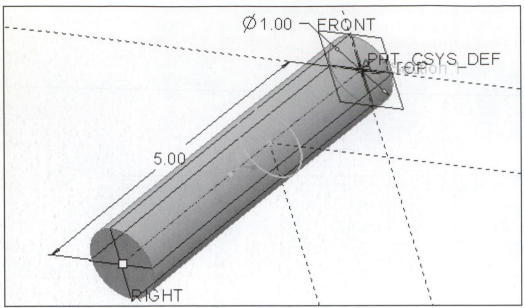

**Figure 1.2** Pin

## Creating the Pin Part

In this lesson most commands and picks will be accompanied by the window or dialog box that will open as the command is initiated. After this lesson, the step-by-step boxes will *not* show every window and dialog box, as this would make the text extremely long. Appropriate illustrations will be provided.

*Throughout the text, a box surrounds all commands and menu selections.*

Open **Creo Parametric 2.0** using a shortcut icon on your Desktop *(or with WINDOWS, click: Start > All Programs > Creo Parametric 2.0)*. Creo Parametric 2.0 will open on your computer. Click: **Select Working Directory**

from the Home tab Ribbon > select *your* working directory [Fig. 1.3(a)] or accept the default directory > **OK** > New- **Create a new model** from the Quick Access Toolbar at the upper left of the screen or from the Home tab > Type ⦿ ☐ Part > Name-- *type* **pin** > ☑ Use default template [Fig. 1.3(b)] > **OK** [Fig. 1.3(c)] *(Do not change directories if you have one assigned as a default by your school or company.)*

**Figure 1.3(a)** Select Working Directory Dialog Box *(your directory will be different)*

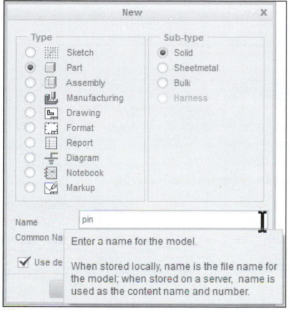

**Figure 1.3(b)** New Dialog Box

**Figure 1.3(c)** Model Tree and Graphics Window *(default datum planes and coordinate system)*

From the keyboard, **Ctrl+S** (saves the object) > **Enter** >  **Extrude** from Model tab Ribbon > **Placement** tab in the Dashboard [Fig. 1.4(a)] > **Define** [Fig. 1.4(b)]

**Figure 1.4(a)** Extrude Placement Tab

**Figure 1.4(b)** Placement Panel, Define an Internal Sketch

Select the **FRONT** datum plane from the Graphics Window (or Model Tree) [Fig. 1.4(c)]. The FRONT datum plane is the Sketch Plane and the RIGHT datum plane is automatically selected as the Sketch Orientation Reference with an Orientation of Right [Fig. 1.4(d)].

**Figure 1.4(c)** Select the FRONT Datum Plane

**Figure 1.4(d)** Sketch Dialog Box

Click: **Sketch** from the Sketch dialog box [Fig. 1.4(e)] >  **Center and Point- Create circle by picking the center and a point on the circle** from the Sketch tab Ribbon [Fig. 1.4(f)]

**Figure 1.4(e)** Activate the Sketcher

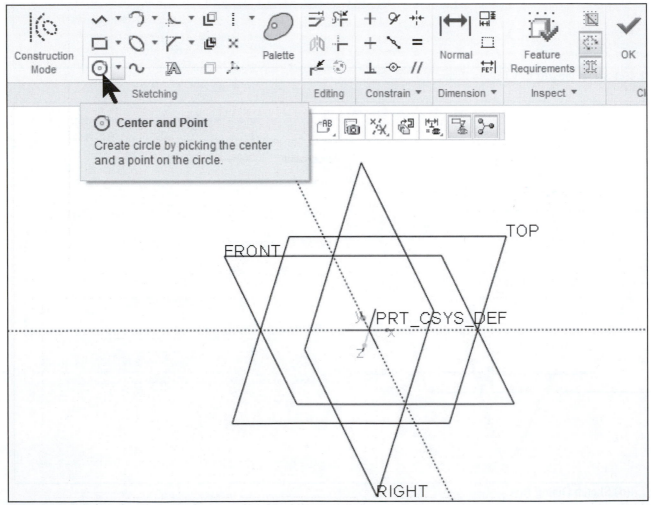

**Figure 1.4(f)** Select the Circle Tool

Pick the origin for the circle's center [Fig. 1.4(g)] > slowly drag the pointer away from the center point and pick a point on the circle's edge [Fig. 1.4(h)] > move the pointer off of the circle's edge > **MMB** (**M**iddle **M**ouse **B**utton) [Figs. 1.4(i-l)] > double-click on the dimension > *type* **1.00** > **Enter**

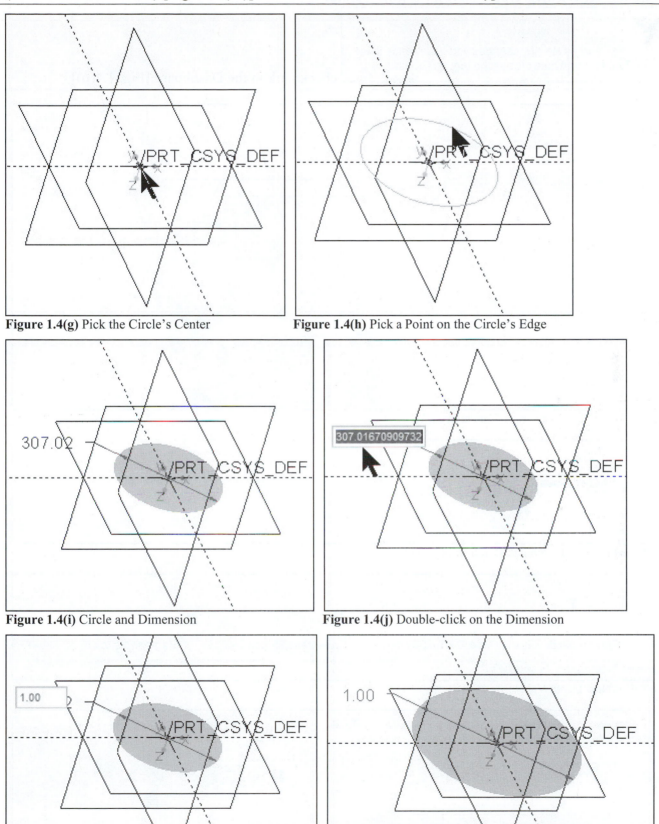

**Figure 1.4(g)** Pick the Circle's Center

**Figure 1.4(h)** Pick a Point on the Circle's Edge

**Figure 1.4(i)** Circle and Dimension

**Figure 1.4(j)** Double-click on the Dimension

**Figure 1.4(k)** Type **1.00**

**Figure 1.4(l)** Completed Sketch

27

Click: **OK- Save the section and exit** (checkmark in the Ribbon) [Fig. 1.4(m)] > click in the depth value field [Fig. 1.4(n)] > *type* **5.00** [Fig. 1.4(o)] > **Enter** >

Applies and saves any changes you have made in the feature tool and then closes the tool dashboard. checkmark in the Dashboard [Fig. 1.4(p)]

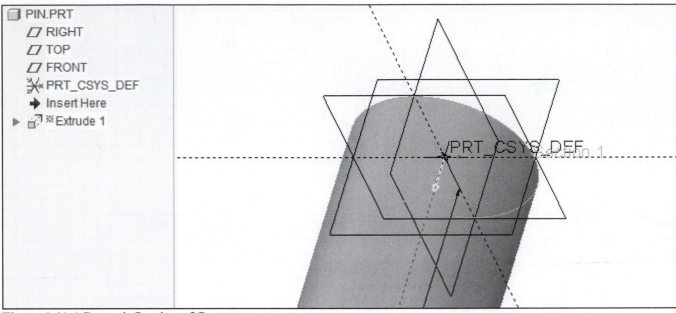

**Figure 1.4(m)** Dynamic Preview of Geometry

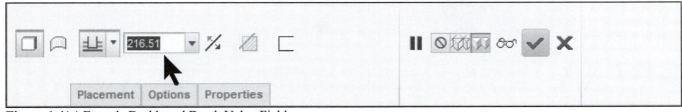

**Figure 1.4(n)** Extrude Dashboard Depth Value Field

**Figure 1.4(o)** New Depth Value

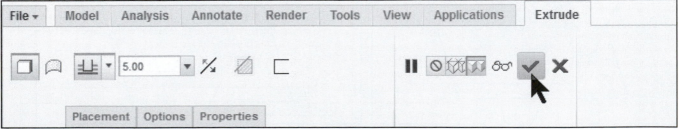

**Figure 1.4(p)** Complete the Extrusion

Click:  **Refit** from the Graphics Toolbar [Figs. 1.4(q-r)] > **File** > **Save** [Fig. 1.4(s)] > **OK** [Fig. 1.4(t)] > **Close** from the Quick Access Toolbar (or File > Close)

**Figure 1.4(q)** Refit

**Figure 1.4(r)** Completed Pin

**Figure 1.4(s)** Save the opened model

**Figure 1.4(t)** Save Object Dialog Box

You have completed your first Creo Parametric 2.0 component.

**Figure 1.5** Plate

## Creating the Plate Part

The second part (Fig. 1.5) will include new commands along with ones used previously in the Pin. Throughout the book each Lesson has repeated and new commands in order to reinforce what you have already learned and adds new capabilities to increase your experience with the software.

Since there are normally 3 to 4 ways of doing almost every command, the book will introduce alternative ways of accomplishing each activity. As an example, to end a command such as Extrude, you can click the **check mark** on the Dashboard, press **Enter** on the keyboard, or click the **MMB (Middle Mouse Button)**.

Click: [Select Working Directory] from the Home tab Ribbon > select *your* working directory [Fig. 1.3(a)] or accept the default directory > **OK**

***(Do not change directories if you have one assigned as a default by your school or company.)***

Click:  **Create a new model** > Type ⊙ ⬚ Part [Fig. 1.6(a)] > Sub-type **Solid** [Fig. 1.6(b)] > Name-- *type* **plate** > ☑ Use default template > click the **MMB** (Middle Mouse Button) *(or press **Enter** or pick **OK**)* [Fig. 1.6(c)] > 🖫 **Save** from the Quick Access Toolbar *(Creo Parametric 2.0 does NOT automatically save for you. So manually save after the completion of each feature.)* > click the **MMB** *(or press **Enter** or pick **OK**)*

**Figure 1.6(a)** New Dialog Box, Part is the Default          **Figure 1.6(b)** Sub-type Solid is the Default

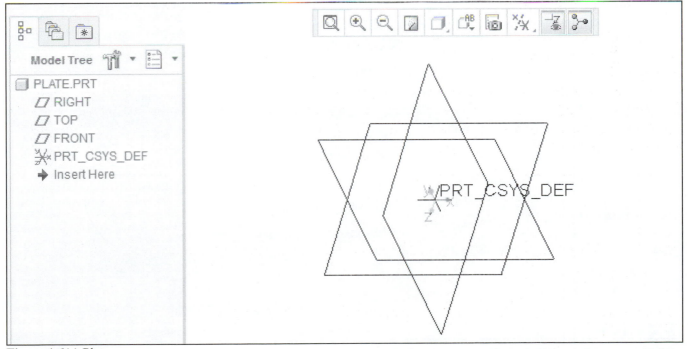

**Figure 1.6(c)** Plate

Click: **File** tab > **Options** [Fig. 1.6(d)] > **Sketcher** > scroll down if necessary > ☑ Show the grid >
☑ Snap to grid [Fig. 1.6(e)] > **OK** > **No**

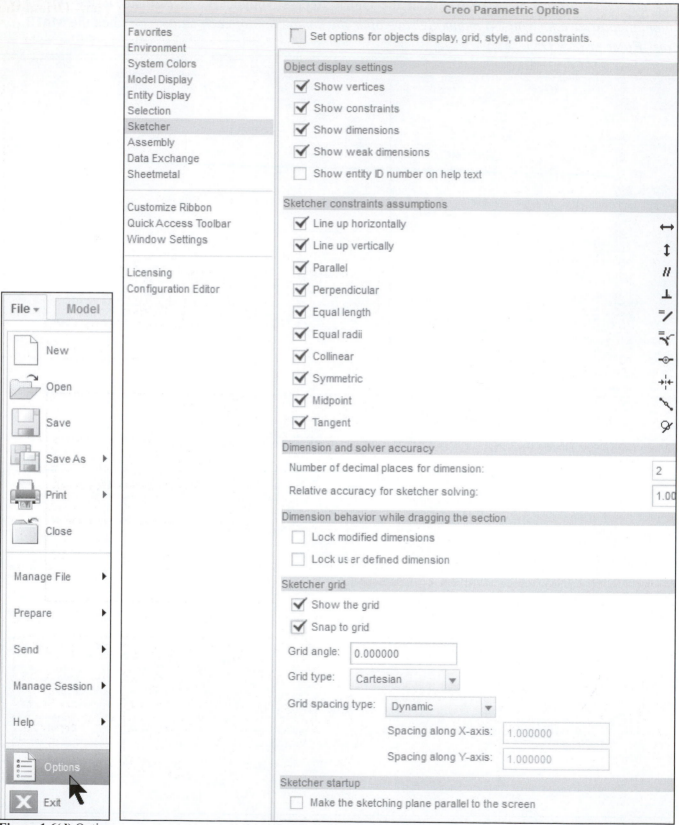

**Figure 1.6(d)** Options

**Figure 1.6(e)** Sketcher Options

Pick the **FRONT** datum plane [Fig. 1.7(a)] >  Extrude from the Model tab Ribbon > from the Graphics Toolbar [Fig. 1.7(b)] or Sketch tab Ribbon Setup Group [Fig. 1.7(c)], click: **Sketch View- Orient the sketching plane parallel to the screen**

**Figure 1.7(a)** Select the Front Datum Plane

**Figure 1.7(b)** Sketch View from the Graphics Toolbar

**Figure 1.7(c)** Sketch View from the Sketch tab Ribbon Setup Group

Click: **View** tab > from the Ribbon, 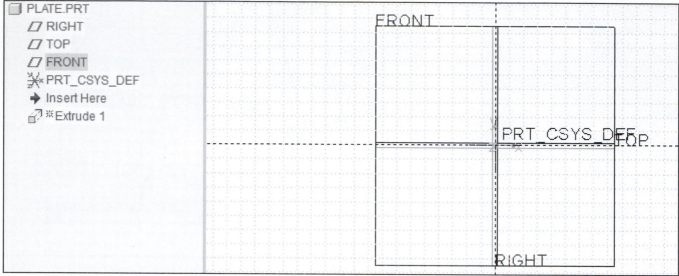 toggle all *on* > **Refit** from the Graphics

Toolbar > **Sketch** tab [Fig. 1.7(d)] > **Import data from a palette into the active object** [Fig. 1.7(e)] > within the Polygons tab (default selection), scroll down to see the octagon > double-click on **Octagon** [Fig. 1.7(f)]

---

PLATE.PRT
- RIGHT
- TOP
- FRONT
- PRT_CSYS_DEF
- Insert Here
- Extrude 1

**Figure 1.7(d)** Sketch Plane parallel to the screen

**Figure 1.7(e)** Sketcher Palette

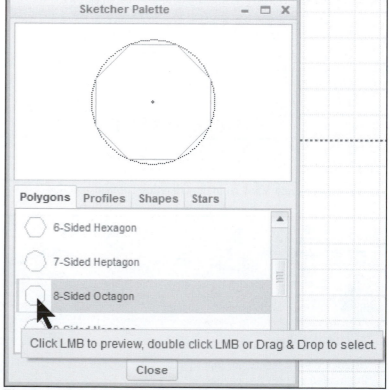

**Figure 1.7(f)** Polygons tab: 8-Sided Octagon

Click: **LMB** to place the octagon on the sketch (by picking a position not on the CYSY) [Fig. 1.7(g)] > place the pointer at the center of the octagon > press and hold down the **LMB** (**L**eft **M**ouse **B**utton) to drag (move) the octagon > drop (release the **LMB** at) the center of the octagon at the origin > modify the Scaling Factor value to **2** [Fig. 1.7(h)] > **Enter** > ✓ in the Ribbon > **Close** from the Sketcher Palette

**Figure 1.7(g)** Place Octagon anywhere on the Sketch by picking a Position with the **LMB**

**Figure 1.7(h)** Drag and Drop the Center of the Octagon at the Origin of the Part (PRT_CSYS_DEF)

Click:  **Refit** [Fig. 1.7(i)] > in the Graphics Window, press and hold the **RMB** (**R**ight **M**ouse **B**utton)

> **OK** > from the Graphics Toolbar > **Standard Orientation** [Fig. 1.7(j)]

> modify the depth value in the Dashboard to **2.00** > **Enter** [Fig. 1.7(k)] > from the Dashboard [Fig. 1.7(l)] > **LMB** to deselect *(place the mouse pointer in the Graphics Window, making sure that the mouse pointer is NOT on an object or within a view boundary; click: **LMB**)* [Fig. 1.7(m)] > > > **OK**

**Figure 1.7(i)** Octagon Sketch

**Figure 1.7(j)** Standard Orientation

**Figure 1.7(k)** Modify Depth to **2.00**

**Figure 1.7(l)** Depth Preview

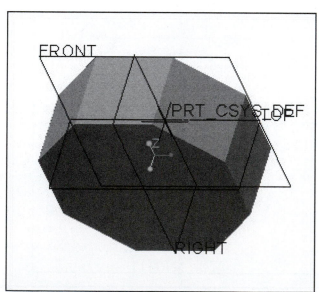

**Figure 1.7(m)** Extruded Octagon

36

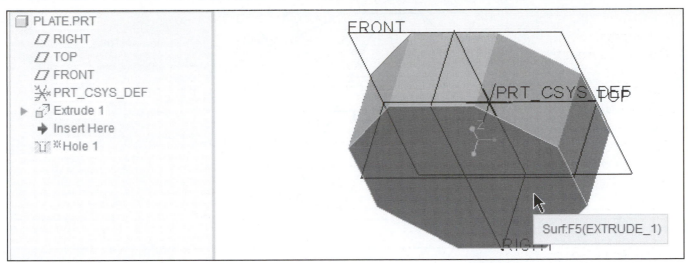

**Figure 1.8(a)** Select the Surface

**Figure 1.8(b)** Hole Previewed

37

In the Graphics Window *(making sure that the pointer is NOT on an object or within a view boundary)*, press and hold: **RMB** [Fig. 1.8(c)] > **Offset References Collector** > press and hold the **Ctrl** key > select datum **RIGHT** from the Model Tree > with the **Ctrl** key still pressed, select datum **TOP** from the Model Tree [Fig. 1.8(d)] > release the **Ctrl** key > **Placement** tab from the Dashboard [Fig. 1.8(e)]

**Figure 1.8(c)** RMB > Offset References Collector

**Figure 1.8(d)** Select the RIGHT and the TOP Datum Planes (from the Model Tree) as Offset References

**Figure 1.8(e)** Hole Dashboard Placement Tab

In the Offset References collector, click on **Offset**, next to RIGHT [Fig. 1.8(f)] > click on **Offset** > **Align** [Fig. 1.8(g)] > click on **Offset**, next to TOP [Fig. 1.8(h)] > click on **Offset** > **Align** [Figs. 1.8(i-j)]

**Figure 1.8(f)** Click on Offset, next to RIGHT

**Figure 1.8(g)** Click Align

**Figure 1.8(h)** Click on Offset, next to TOP

**Figure 1.8(i)** Click Align

**Figure 1.8(j)** Hole Preview

Click: 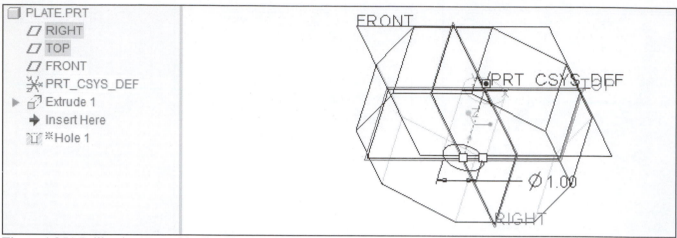 > **Drill to intersect with all surfaces** [Fig. 1.8(k)] > modify the hole's diameter, *type* **1.00** [Fig. 1.8(l)] > **Enter** > **Placement** tab *(to close the Placement Panel)* > from the Dashboard > > > **Display Style** from the Graphics Toolbar > Shading With Edges > **LMB** to deselect [Fig. 1.8(m)] > **Ctrl+S** from the keyboard to save > **OK**

**Figure 1.8(k)** Drill to intersect with all surfaces

**Figure 1.8(l)** Diameter **1.00**

**Figure 1.8(m)** Completed Hole

Click: **View** tab > Appearance Gallery [Fig. 1.9(a)] > ● **ptc-metallic-gold** [Fig. 1.9(b)] > 🖌 > click on the object name in the Model Tree [Model Tree ... PLATE.PRT / RIGHT] > [Select — Select 1 or more items. OK Cancel] **OK** > Refit from the Ribbon > [▼ 📐 Extrude 1 / 📝 Section 1] in the Model Tree > 💾 **Save** from the Quick Access Toolbar > **Enter** (Fig. 1.10) > **File** > **Manage Session** > **Erase Current** > **Yes**

**Figure 1.9(a)** ptc-metallic-gold

**Figure 1.9(b)** Appearance Gallery

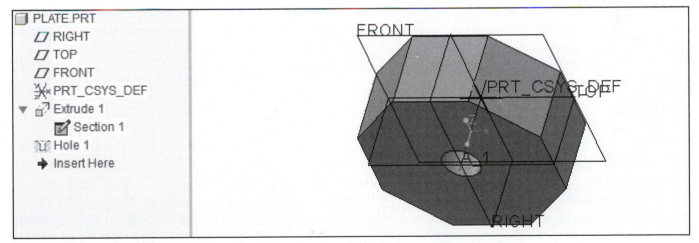

**Figure 1.10** Completed Plate

　　You have completed your second Creo Parametric 2.0 component. Next, you will put the parts together to create an assembly.

**Figure 1.11** Assembly

## Creating the Assembly

Using the two parts just modeled you will now create an assembly (Fig. 1.11). Just as you can combine features into parts, you can also combine parts into assemblies. The Assembly mode in Creo Parametric 2.0 enables you to place component parts and subassemblies together to form assemblies.

As with a part, an assembly starts with default datum planes and a default coordinate system. To create a subassembly or an assembly, you must first create datum features. You then create or assemble additional components to the existing component(s) and datum features.

One new addition to PTC's software is the **3D Dragger**.

The 3D Dragger is a graphical tool for precision handling of assembly components. The *axes* are called **draggers**. Pull or rotate the draggers to make changes to the components position and orientation. The draggers enable movement in one or more *degrees of freedom* (DOFs). They support linear, planar, free, translation, and angular movements. You can pull a dragger along a 2D or 3D trajectory. When you move a geometric entity using a dragger, the relative distances from the start point of the entity appear in the graphics window. You can hold down the SHIFT key while pulling a dragger to snap the dragger to a geometric reference.

Click:  **Create a new model** > Type [Assembly] > Sub-type **Design** > Name-- *type* **connector** > ☑ Use default template [Fig. 1.12(a)] > **OK** > **View** tab from the Ribbon > Display Style ▾ toggle all *on* > **Settings** from the Navigator [Fig. 1.12(b)] > Tree Filters... > toggle all options *on* [Fig. 1.12(c)] > **Apply** > **OK** [Fig. 1.12(d)] > **Refit** > **File** > **Save** > **OK**

**Figure 1.12(a)** New Dialog Box, Assembly

**Figure 1.12(b)** Tree Filters

**Figure 1.12(c)** Model Tree Items Dialog Box

**Figure 1.12(d)** Assembly Default Datums and Coordinate System

Click: **Model** tab > [Assemble] > File name: **plate.prt** [Fig. 1.13(a)] > **Open**

**Figure 1.13(a)** Open Dialog Box, Type Creo Files

Click: 🔍 **Refit** from the Graphics Toolbar > [Automatic] [Fig. 1.13(b)] > [Default] [Fig. 1.13(c)] > ✔ from the Dashboard

**Figure 1.13(b)** Assembly Dashboard

**Figure 1.13(c)** Default Constraint

Click: **Ctrl+D** [Fig. 1.13(d)] > **Ctrl+S** > **OK** > in the Model Tree, expand  > expand

> expand > **View** tab > from the Ribbon,

toggle all *off* > **Model** tab > > Shading With Edges [Fig. 1.13(e)]

**Figure 1.13(d)** Fully Constrained Component at the Default Position *(coordinate system to coordinate system)*

**Figure 1.13(e)** CONNECTOR Assembly

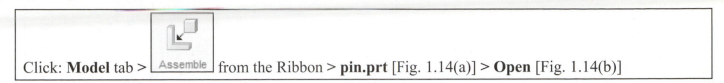

Click: **Model** tab > [Assemble icon] from the Ribbon > **pin.prt** [Fig. 1.14(a)] > **Open** [Fig. 1.14(b)]

**Figure 1.14(a)** Select pin.prt

**Figure 1.14(b)** Component Shown in Assembly Window with No Constraints and 3D Dragger Displayed

Click: **Datum Display Filters > Select All** (unchecked) > **LMB** in the Graphics Window >  **Hide 3D Dragger** from the Dashboard > pick on the Pin's cylindrical surface [Fig. 1.14(c)] > pick on the Plate's cylindrical hole surface [Fig. 1.14(d)]

**Figure 1.14(c)** Select the Pin's Cylindrical Surface

**Figure 1.14(d)** Select the Plate's Cylindrical Hole Surface

Click: **Placement** tab > **New Constraint** [Fig. 1.14(e)]

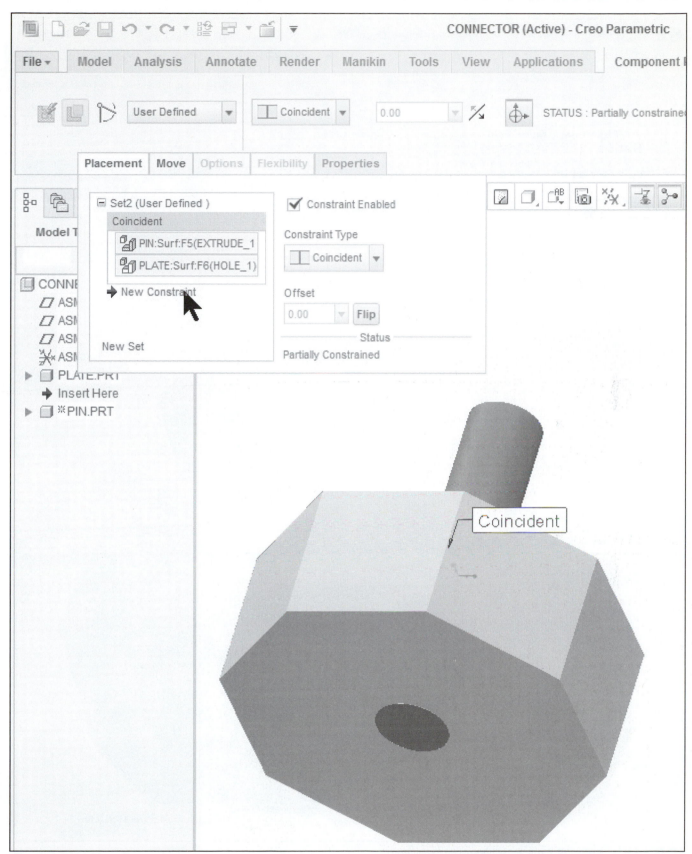

**Figure 1.14(e)** New Constraint

Pick on the Plate's surface [Fig. 1.14(f)]

**Figure 1.14(f)** Select the Plate's Surface

Pick on the Pin's end face surface [Fig. 1.14(g)]

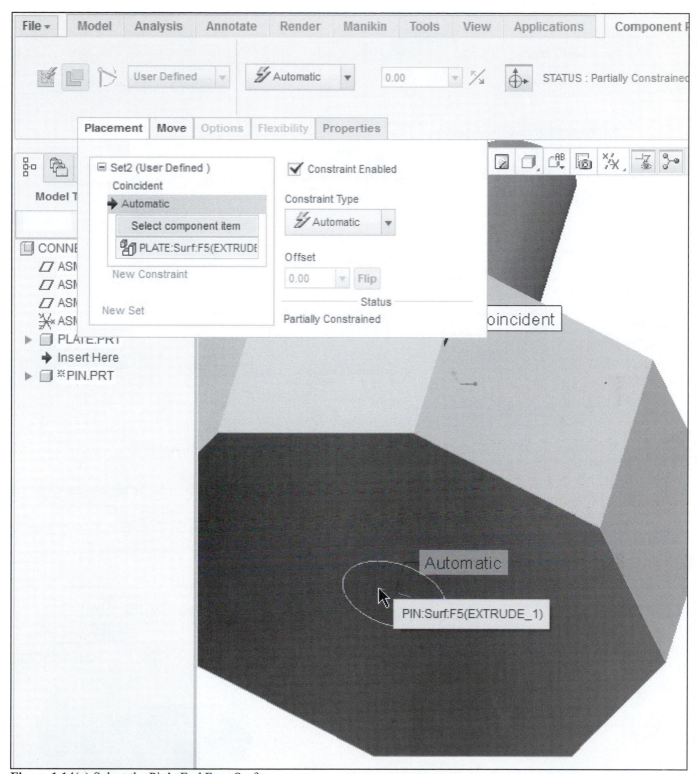

**Figure 1.14(g)** Select the Pin's End Face Surface

Click: **Coincident** [Fig. 1.14(h)] > **Distance** [Fig. 1.14(i)] > click in the Offset value field, *type* **.50** [Fig. 1.14(j)] > **Enter**

**Figure 1.14(h)** Select Coincident

**Figure 1.14(i)** Changing Constraint Type to Distance

**Figure 1.14(j)** Offset .50 *(if your offset is in the opposite direction, enter a value of -.50)*

Click: [✓] from the Dashboard [Fig. 1.14(k)] > [▼] expand all in Model Tree > [X̶X̶] **Datum Display Filters** from the Graphics Toolbar > [✓ (Select All)] [Fig. 1.14(l)] > **LMB** in the Graphics Window > [🔍] **Refit** > **Ctrl+S** > **OK**

**Figure 1.14(k)** Assembly Dashboard: Fully Constrained

**Figure 1.14(l)** Assembled Pin

Click: **View** tab >  >  *off* > **File** tab >

**Manage File** >  >  [Fig. 1.14(m)] > **File** > **Save** >

**Enter** > **File** > **Save As** > Type  > **Zip File (*.zip)** *(scroll to the end of the list)* > **OK** > upload the zip file to your course interface or attach to an email and send to your instructor and/or yourself *(your zip files are small in size, so this is also a good way to back up your work online)* > **File** > **Close**

**Figure 1.14(m)** Assembled Pin *(the quality of your graphics card and graphics settings may prevent this display)*

You have completed your first Creo Parametric 2.0 assembly. Next you will create a set of drawings for the Connector assembly and the Plate component.

**Figure 1.15** Assembly Drawing and Detail Drawing

## Creating Drawings

Using the parts and the assembly just created, you will now create an assembly drawing of the Connector and a detail drawing of the Plate (Fig. 1.15).

The Drawing mode in Creo Parametric 2.0 enables you to create and manipulate engineering drawings that use the 3D model (part or assembly) as a geometry source. You can pass dimensions, notes, and other elements of design between the 3D model and its views on a drawing.

Click:  **New** > Type ⦿ 🖳 Drawing > Name-- *type* **connector** > ☑ Use default template [Fig. 1.16(a)] > **OK** [Fig. 1.16(b)] > Default Model **Browse** > **Preview** *on* > **connector.asm** [Fig. 1.16(c)]

**Figure 1.16(a)** New Dialog Box, Drawing

**Figure 1.16(b)** New Drawing Dialog Box

**Figure 1.16(c)** Connector Assembly

Click: **Open** > **OK** from the New Drawing dialog box [Fig. 1.16(d)] > **Layout** tab > in the Graphics Window, press and hold: **RMB** > **Sheet Setup** from the short cut menu [Fig. 1.16(e)]

**Figure 1.16(d)** Drawing

**Figure 1.16(e)** Drawing Sheet Setup

Sheet Setup dialog box displays [Fig. 1.16(f)] > click in the **C Size** field [Fig. 1.16(g)] > to see options, click: [Fig. 1.16(h)] > **Browse** [Fig. 1.16(i)] > **c.frm** [Fig. 1.16(j)] > **Open** [Fig. 1.16(k)] > **OK**

**Figure 1.16(f)** Sheet Setup Dialog Box

**Figure 1.16(g)** C Size Field

**Figure 1.16(h)** Format Options

**Figure 1.16(i)** Browse

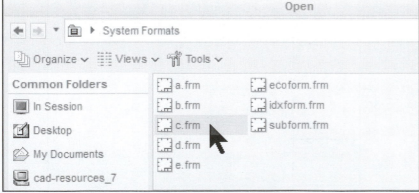

**Figure 1.16(j)** c.frm from the System Formats Library

**Figure 1.16(k)** Format C

In the Graphics Window, press and hold: **RMB > Insert General View** [Fig. 1.16(l)] > **No Combined State** *(selected by default)* [Fig. 1.16(m)] > **OK** > pick a position for the new view [Fig. 1.16(n)]

**Figure 1.16(l)** Insert General View

**Figure 1.16(m)** Select Combined State Dialog Box **Figure 1.16(n)** Select a Position for the New View

59

Click: **OK** [Fig. 1.16(o)] >  **Refit** > **LMB** (in the Graphics Window to deselect) > **File** > **Save** > **OK** [Fig. 1.16(p)] > **File** > **Save As** > Type ▼ > **Zip File (*.zip)** *(scroll to the end of the list)* > **OK** > upload to your course interface or attach to an email and send to your instructor and/or yourself > **File** > **Close**

**Figure 1.16(o)** Drawing View Dialog Box

**Figure 1.16(p)** Completed Assembly Drawing

Click:  **New** > Type ● 🖳 Drawing > Name-- *type* **plate** > ☑ Use default template [Fig. 1.17(a)] > **OK** [Fig. 1.17(b)] > Default Model **Browse** > **Preview** *on* > **plate.prt** [Fig. 1.17(c)] > **Open**

**Figure 1.17(a)** New Dialog Box, Drawing          **Figure 1.17(b)** Browse

**Figure 1.17(c)** Plate Preview

Click: **OK** [Fig. 1.17(d)] drawing opens > in the Graphics Window, press and hold: **RMB** > **Sheet Setup** [Figs. 1.17(e-f)] > click in the **C Size** field [Fig. 1.17(g)] > ⏷ to see options [Fig. 1.17(h)]

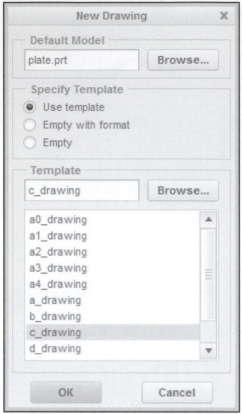

**Figure 1.17(d)** New Drawing Dialog Box

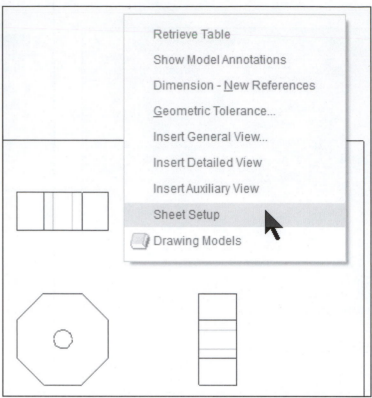

**Figure 1.17(e)** Plate Drawing, RMB > Sheet Setup

**Figure 1.17(f)** Sheet Setup Dialog Box

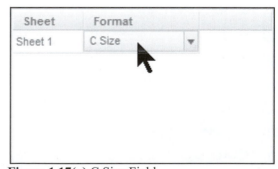

**Figure 1.17(g)** C Size Field

Click: **Browse** [Fig. 1.17(i)] > double-click on **c.frm** [Fig. 1.17(j)] > **OK** formatted drawing is displayed [Fig. 1.17(k)] > **Ctrl+S** > **Enter**

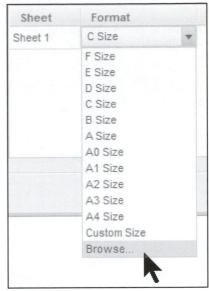

**Figure 1.17(h)** Sheet Setup Dialog Box, Format Options

**Figure 1.17(i)** Browse

**Figure 1.17(j)** c.frm

**Figure 1.17(k)** Formatted Drawing

In the Graphics Window, press and hold: **RMB > Insert General View** [Fig. 1.17(l)] > **No Combined State** *(selected by default)* > **OK** > pick a position for the new view [Fig. 1.17(m)] > **OK** from the Drawing View dialog box [Fig. 1.17(n)] >  **Refit**

**Figure 1.17(l)** Insert General View

**Figure 1.17(m)** Select a Position for the New View

**Figure 1.17(n)** Drawing View Dialog Box

In the Graphics Window, press and hold: **RMB > Lock View Movement** *off (unchecked)* [Fig. 1.17(o)]

**Figure 1.17(o)** Lock View Movement off

Place the pointer on the *selected* view > press and hold down the **LMB** > move the pointer as desired [Fig. 1.17(p)] > release the **LMB** > **LMB** in the Graphics Window to deselect the view [Fig. 1.17(q)] > **Ctrl+S** > **Enter** > **File** > **Save As** > Type  > **Zip File (*.zip)** > **OK** > upload to your course interface or attach to an email and send to your instructor or yourself > **File** > **Manage File** > **Delete Old Versions** > **Enter** > **File** > **Close** > **File** > **Manage Session** > **Erase Not Displayed** > **Enter** > **File** > **Exit** > **Yes**

**Figure 1.17(p)** Move the View

**Figure 1.17(q)** Deselect the View

You have completed the Creo Parametric 2.0 overview. Spend some time exploring the website ***www.cad-resources.com*** for downloads, materials, and other CAD related information.

**Figure 2.1** Part T-Nut

## OBJECTIVES

- Understand and customize the **User Interface (UI)**
- **Download 3D CAD models** from the Creo Standard Parts Library (old Pro/Library)
- Experience the **Creo Parametric 2.0 Browser**
- Master the **File Functions**
- Become familiar with the **Help** center
- Use the **Command Locator**
- Be introduced to the **Display** and **View** capabilities
- Use **Mouse Buttons** to **Pan**, **Zoom**, and **Rotate** the object
- Change **Display Settings**
- Investigate an object with **Information Tools**
- Experience the **Model Tree** functionality

## REFERENCES AND RESOURCES

For **Resources** go to **www.cad-resources.com** > click on the Creo Parametric 2.0 book cover

- Lesson 2 Lecture
- Lesson 2 3D models embedded in a PDF
- Creo Parametric 2.0 Quick Reference Card
- Creo Parametric 2.0 Configuration Options

# Creo Parametric 2.0

This lesson will introduce you to the Creo Parametric 2.0 working environment. An existing 3D CAD model (Fig. 2.1) from the Creo Standard Parts Library will be used to demonstrate the UI (user interface) and the general interaction required to master this CAD software.

## Creo Parametric 2.0 Interface

The Creo Parametric 2.0 interface consists of a navigation window, an embedded Web browser, toolbars, information areas, and the graphics window (Fig. 2.2). Each Creo Parametric 2.0 object opens in its own window. You can perform many operations from the Ribbon in multiple windows without canceling pending operations. Only one window is active at a time, but you can still perform some functions in the inactive windows. To activate a window simply press Ctrl+A.

**Figure 2.2** Creo Parametric 2.0 Interface

The Creo Parametric 2.0 window consists of the following elements:

- **Creo Parametric 2.0 Browser** An embedded Web browser is located to the right of the navigator window. This browser provides you access to internal or external Web sites.

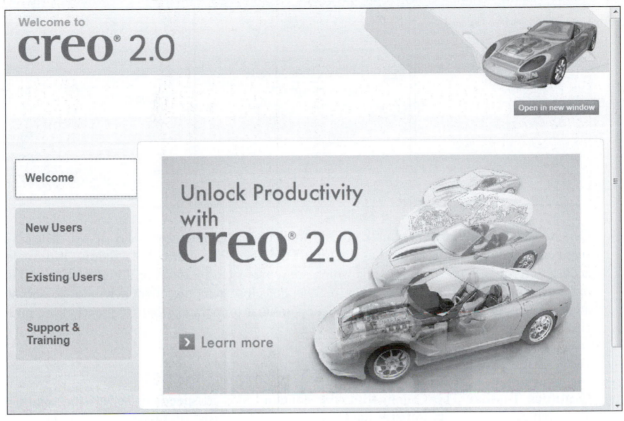

- **Graphics Window** The graphics window is the main working space (main window) for modeling.

- **Navigation Area** The navigation area is located on the left side of the Graphics Window. It includes tabs for the Model Tree and Layer Tree, Folder Browser, Favorites, History and Connections.  on the status bar/message area (lower left corner of the Creo Parametric 2.0 window) controls the display of the navigator.

- **Quick Access Toolbar** The Quick Access toolbar is available regardless of which Ribbon tab is selected. By default it is located at the top of the Creo Parametric 2.0 window. It provides quick access to frequently used commands which are represented by icons or buttons, such as buttons for opening and saving files, undo, redo, regenerate, close window, switch windows, and so on. In addition, you can customize the Quick Access toolbar to include other frequently used buttons and cascading lists.

- **Graphics Toolbar** The Graphics toolbar contain icons to speed up access to commonly used View tab commands. By default, the Graphics toolbar consist of a row of buttons located directly under the Ribbon. The Graphics toolbar can be positioned on the top, left, bottom, and right of the graphics window, or in the status bar, or turned off. Graphics toolbar buttons can be added or removed to customize the layout.

- **File Menu** The File button in the upper-left corner of the Creo Parametric 2.0 window opens a menu that contains commands for managing files, models, preparing models for distribution, and for setting your Creo Parametric 2.0 environment and configuration options.

- **Ribbon** The Ribbon contains command buttons organized in a set of tabs. On each tab, the related buttons are grouped. You can minimize the Ribbon to make more space available on your screen. You can customize the Ribbon by adding, removing, or moving buttons. The following figure shows the different elements of a Ribbon.

- **Tabs** Tabs are usually available when you are in a particular mode or application, or when you need them in a particular context. The Model, Analysis, Annotate, Render, Tools, View, Applications tabs are some commonly available tabs when you are in a mode. The Home tab is available when Creo Parametric 2.0 has no model opened. Tabs related to a particular context open or close automatically when you activate or deactivate the context. Similarly, tabs related with a particular object open or close automatically when the related objects are selected or deselected, respectively. Tabs that contain tools of an application or controls of a tool have specific buttons to open and close them.

**1. Tab  2. Button  3. Group  4. Group overflow button  5. Dialog box launcher**

- **Screen Tips** The status bar messages also appear in small boxes near the menu option or dialog box item or toolbar button that the mouse pointer is passing over.

- **Message Area** Each Creo Parametric 2.0 window has a status bar/message area located at the bottom of the Creo Parametric 2.0 window.

  - PRT0001 regeneration completed successfully.
  ⇨ Select a sketch. (If an internal sketch is preferred, the "Define" option can be found in the
  ⇨ Select the center of a circle.
  - When a constraint is active, right click to cycle through lock/disable/enable the constraint.
  ⇨ Select a point on the arc/circle.
  - When a constraint is active, right click to cycle through lock/disable/enable the constraint.

- **Command Locator** The Command Locator will quickly locate and initiate any command by opening the Ribbon and group associated with the command and listing optional locations.

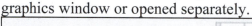

- **Accessory Window** When you place a component or use a tool that includes an accessory component or feature, you can open the accessory component and its Model Tree in a new window, even if it is already open in another window. An accessory window can be docked in the graphics window or opened separately.

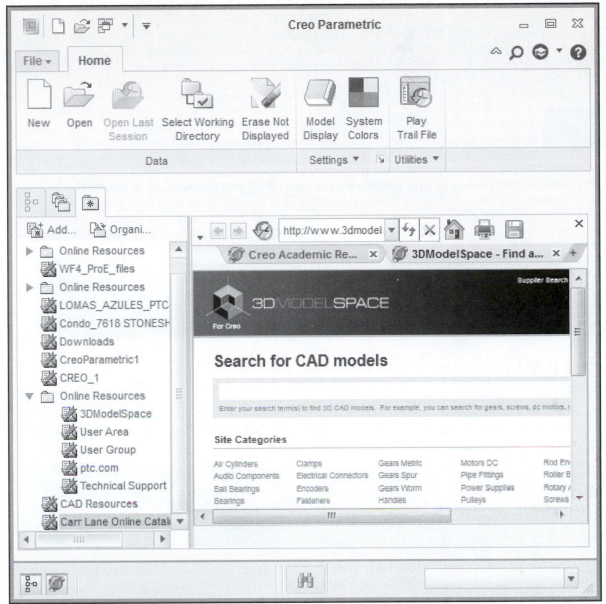

**Figure 2.3** 3DMODELSPACE Catalog

## Catalog Parts

In order to see and use the Creo Parametric 2.0 UI, we must have an active object. You will download an existing 3D standard component using the Browser (Fig. 2.3). You will also experience using the Catalog of parts from PTC. You can access online catalogs from the **3DModelSpace** link. 3DModelSpace is available as a PTC Resource link in the Online Resources folder of Favorites tab in the Navigation area. The home page for the catalog lists catalogs of standard components.

You can browse catalogs, search for a component, view the details of the components, and download the selected component in Creo Parametric 2.0. You can also add a shortcut to a catalog. To access the online component catalogs, click the **Favorites** tab and expand the Personal Favorites folder. Online Resources is located within the Personal Favorites folder. Expand the Online Resources folder and select 3DModelSpace. The catalogs page opens in the Creo Parametric 2.0 browser and you can select a catalog to browse or you can right-click on a catalog to add a shortcut to it to be accessed from the Online Resources folder. You can also add the catalog shortcut to **Personal Favorites**.

**Note: The PTC website may change names of certain items as they update their product library.**

Launch **Creo Parametric 2.0** > **File** > **Manage Session** > **Select Working Directory** > select a directory

for your projects > **OK** >  [Fig. 2.4(a)] >

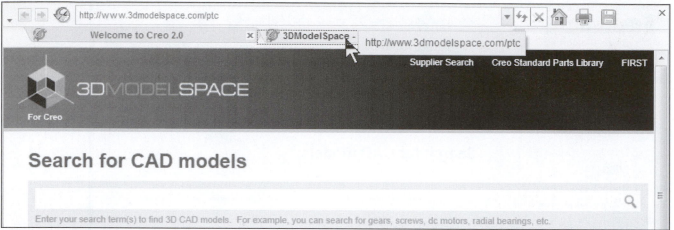 [Fig. 2.4(b)] > [Fig. 2.4(c)]

*(Do not change directories if you have one assigned as a default by your school or company.)*

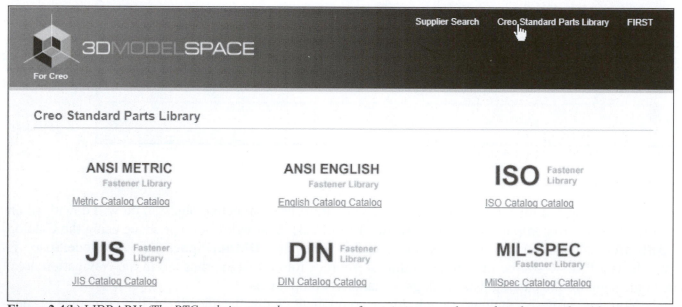

**Figure 2.4(a)** 3DModelSpace

**Figure 2.4(b)** LIBRARY *(The PTC website may change names of certain items as they update their product library)*

**Figure 2.4(c)** ANSI ENGLISH Fastener Library

Scroll to the bottom of the PTC Catalog Directory page in the Browser window and click:

I ACCEPT the Terms and Conditions Stated Above. [Fig. 2.4(d)] > Show All Categories [Fig. 2.4(e)]

**Figure 2.4(d)** ENG-PART LIBRARY Category Images

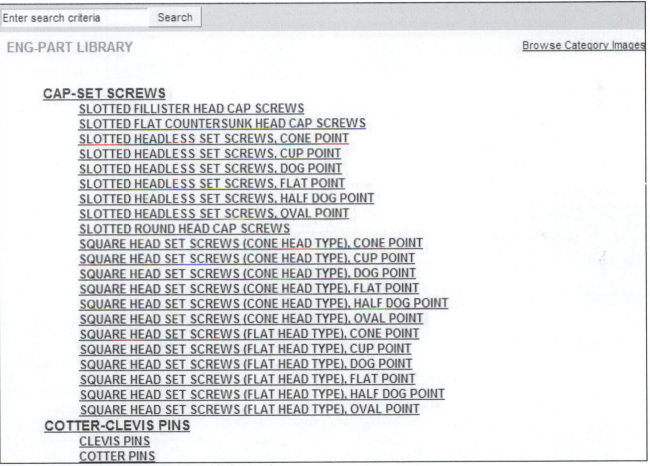

**Figure 2.4(e)** ENG-PART LIBRARY All Categories

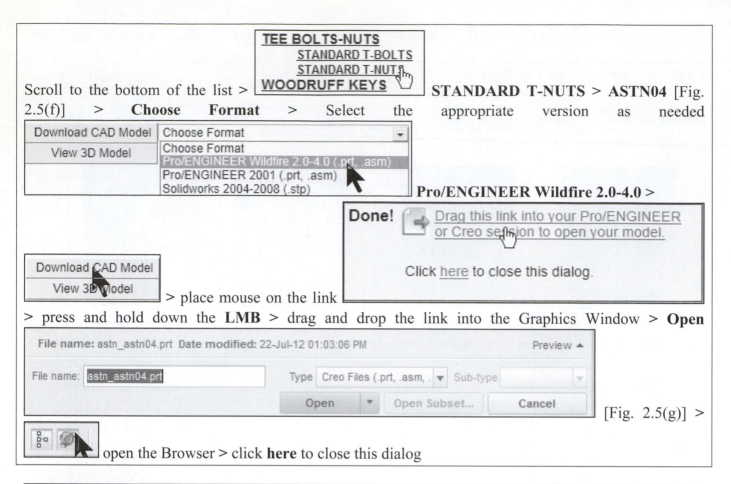

Scroll to the bottom of the list > **STANDARD T-NUTS > ASTN04** [Fig. 2.5(f)] > **Choose Format** > Select the appropriate version as needed **Pro/ENGINEER Wildfire 2.0-4.0 >** > place mouse on the link > press and hold down the **LMB** > drag and drop the link into the Graphics Window > **Open** [Fig. 2.5(g)] > open the Browser > click **here** to close this dialog

| Product Number | General Description | Manufacturer Name | Thickness (9) | Nut Height (9) | Tongue Width (9) | Nut Width (9) | Thread Size | Nominal Diameter Thread Pitch |
|---|---|---|---|---|---|---|---|---|
| | Standard T-Nuts - ANSI/ASME B5.1M-1985 | PTC | Select | Select | Select | Select | Select | Select |
| | Reset          Sort | Reset   Sort | Reset  Sort | Reset  Sort | Reset  Sort | Reset  Sort | Reset  Sort | Reset  Sort |
| ASTN01 | Standard T-Nuts - ANSI/ASME B5.1M-1985 | PTC | 0.281 | 0.188 | 0.330 | 0.562 | 0.312 | 0.250-20 |
| ASTN02 | Standard T-Nuts - ANSI/ASME B5.1M-1985 | PTC | 0.375 | 0.250 | 0.418 | 0.688 | 0.375 | 0.312-18 |
| ASTN03 | Standard T-Nuts - ANSI/ASME B5.1M-1985 | PTC | 0.531 | 0.312 | 0.543 | 0.875 | 0.5 | 0.375-16 |
| ASTN04 | Standard T-Nuts - ANSI/ASME B5.1M-1985 | PTC | 0.625 | 0.406 | 0.668 | 1.125 | 0.625 | 0.500-13 |
| ASTN05 | Standard T-Nuts - ANSI/ASME B5.1M-1985 | PTC | 0.781 | 0.531 | 0.783 | 1.312 | 0.75 | 0.625-11 |
| ASTN06 | Standard T-Nuts - ANSI/ASME B5.1M-1985 | PTC | 1.000 | 0.688 | 1.033 | 1.688 | 1 | 0.750-10 |
| ASTN07 | Standard T-Nuts - ANSI/ASME B5.1M-1985 | PTC | 1.312 | 0.938 | 1.273 | 2.062 | 1.25 | 1.000-8 |
| ASTN08 | Standard T-Nuts - ANSI/ASME B5.1M-1985 | PTC | 1.625 | 1.188 | 1.523 | 2.500 | 1.5 | 1.250-7 |

**Figure 2.5(f)** ENG-PART LIBRARY > T-BOLTS-NUTS > STANDARD T-NUTS

Figure 2.5(g) Drag and Drop the Link *(which is a Zip file of a ProE part)* into the Graphics Window - Library Part

*ALTERNATE MODEL-* *If your system can not access the Creo Standard Parts Library; in the embedded web Browser [Internet Explorer (IE) in this case], type:* ***www.cad-resources.com*** *>*

*click on the appropriate book cover icon > place your mouse cursor on the part icon > drag and drop the link into the Graphics Window >* ***OR*** *> click on the part icon > Save [Fig. 2.5(h)] > Save (if needed) > select the desired folder to save the zip file to > Save > Close (if needed) > Open a file manager window to the desired folder > drag and drop the zip file into your Graphics Window > Open (if needed) >* ***OR*** *> to the saved zip file, unzip or extract all > drag and drop the ProE file into the Graphics Window*

Figure 2.5(h) Zip File Download dialog box – *your dialog box may appear differently*

Click:  close the Browser [Fig. 2.5(i)] > to expand the Model Tree features and Footer [Fig. 2.5(j)] > **RMB** on Interfaces **INTFC001** [Fig. 2.5(k)] > **Info** > **Feature** > close the Browser

**Figure 2.5(i)** Close the PTC Browser

**Figure 2.5(j)** Expand Model Tree

**Figure 2.5(k)** INTERFACES

Click: **File > Save As >** New Name-- *type* **T-NUT** [Fig. 2.5(l)] **> OK > File > Close > Ctrl+O** open an existing model **> t-nut.prt** [Fig. 2.5(m)] **> Open >** to expand Model Tree items [Fig. 2.5(n)]

| Model Name | ASTN04.PRT |
|---|---|
| New Name | T-NUT |
| Type | Part (*.prt) |

OK          Cancel

**Figure 2.5(l)** Save As T-NUT

**Figure 2.5(m)** File Open Dialog Box, **t-nut.prt**

**Figure 2.5(n) t-nut.prt**

Click: **View** tab >  > **ptc-std-aluminum** [Fig. 2.5(o)] > **ptc-std-aluminum** [Fig. 2.5(p)] > > select **T-NUT.PRT** in the Model Tree [Fig. 2.5(q)] > **OK** [Fig. 2.5(r)]

**Figure 2.5(o)** Appearance Gallery

**Figure 2.5(p)** ptc-std-aluminum

Model Tree

T-NUT.PRT
 DTM1
 DTM2
 DTM3
 Protrusion id 39
  Section 1
 Hole id 76
 Chamfer id 96
 Round id 127
 Chamfer id 146
 Insert Here
 Footer

**Figure 2.5(q)** T-NUT

T-NUT.PRT
 DTM1
 DTM2
 DTM3
 Protrusion id 39
 Hole id 76
 Chamfer id 96
 Round id 127

**Figure 2.5(r)** New Color

# File Functions

The **File** tab provides options for opening, creating, saving, renaming, backing up files, and printing. Before using any File tool, make sure you have set your working directory to the folder where you wish to save objects for the project on which you are working. The Working Directory is a directory that you set up to contain Creo Parametric 2.0 files. You can save a Creo Parametric 2.0 file using **Save** [Fig. 2.6(a)] or **Save As** from the File tab. You can save files, make copies of files, or backups of files after creating them. The default folder to save or back up a file is determined as follows: The **My Documents** folder if you have not set a Working Directory. The **Working Directory** you have set for your current session. The folder you last accessed to open, save, save a copy, or back up your file. Each time you save a file, it creates a new version; *nut.prt.1*, *nut.prt.2*, etc. Save fifty times and you have fifty more files; *nut.prt.50*. To purge old versions: **File > Manage File > Delete Old Versions > Enter**. The last (highest number-last saved) version will be the remaining file and the one that opens the next time you access the file.

   **Save As** [Fig. 2.6(b)] is the exportation of Creo Parametric 2.0 files to different formats, and to save files as images. Since the name already exists in session, you cannot save or Rename a file using the same name as the original file name, even if you save the file in a different directory. Creo Parametric 2.0 forces you to enter a unique file name by displaying the message: *An object with that name exists in session. Choose a different name.* Creo Parametric 2.0 file names are restricted to a maximum of 31 characters and have no spaces. A file can be a part, assembly, or drawing. Each is considered an "object" or "model".

**Figure 2.6(a)** Save

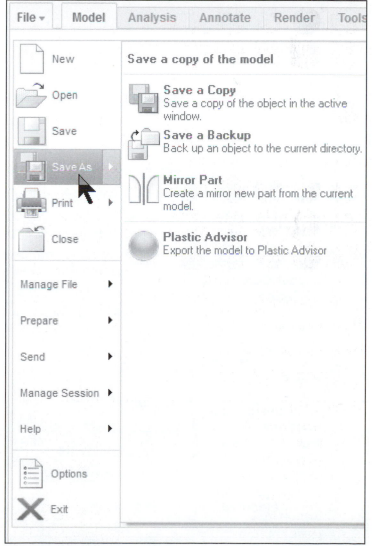

**Figure 2.6(b)** Save As

81

## Help Center

Accessing the 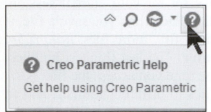 Help function [Fig. 2.7(a)] is one of the best ways to learn any CAD software. Use the Help Center [Fig. 2.7(b)] as often as possible to understand the tool or command you are using at the time and to expand your knowledge of the capabilities provided by Creo Parametric 2.0. Use the **Help** menu to gain access to online information, Creo Parametric 2.0 release information, and customer service information.

**Figure 2.7(a)** Activate the Help Center

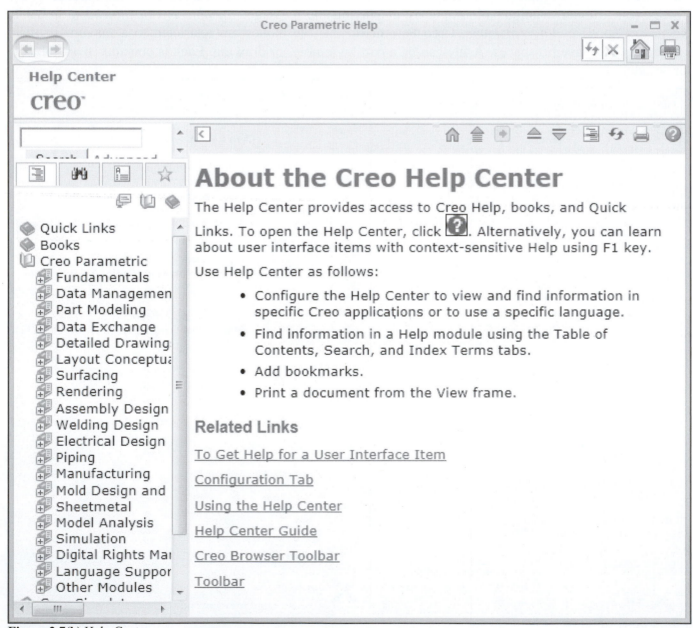

**Figure 2.7(b)** Help Center

For **Context sensitive help**, move your pointer over the icon, dialog box, or menu command for which you want help and press the **F1** key on the keyboard. Click: **Model** tab > move the mouse pointer over

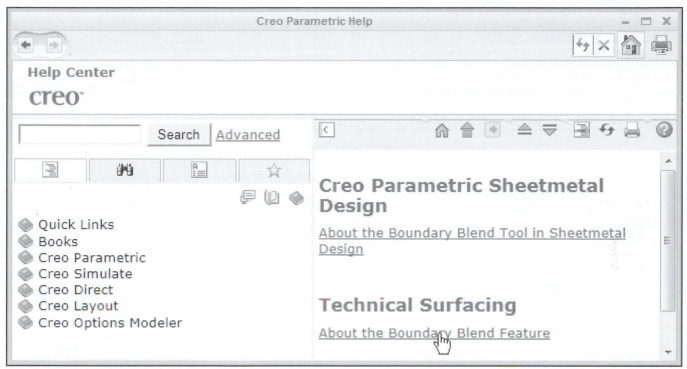 > press **F1** [Fig. 2.8(a)] > **About the Boundary Blend Feature** [Fig. 2.8(b)] > [×] close the Help Center window > repeat the process to investigate other commands

**Figure 2.8(a)** Creo Parametric Help

**Figure 2.8(b)** About the Boundary Blend Feature

## Command Locator

The command search tool enables you to find known commands faster and preview the location of the command on the user interface. You can preview the location only if the command is located on the Ribbon, Quick Access toolbar, Graphics toolbar, or File menu. You can also run a command by clicking the command in the search list.

The tool displays commands not in the Ribbon under the **Commands not in the ribbon** category in the search list. The tool searches commands within the current mode.

Click: 🔍 **Command Search** in the upper top right-hand side of your **screen** [Fig. 2.9(a)] > [_____] Command Locator opens > *type* **View** [View] > [Fig. 2.9(b)] > place pointer on **Perspective View** [Fig. 2.9(c)] > select [Perspective View] [Fig. 2.9(d)] > **Perspective View** in the Render tab Ribbon (toggle *off*) > 🔍 toggle Command Locator shut

**Figure 2.9(a)** Command Search

**Figure 2.9(b)** View Commands

**Figure 2.9(c)** Perspective View

**Figure 2.9(d)** Perspective View

# View and Display Functions

Using the Creo Parametric 2.0 **View** tab [Fig. 2.10(a)], you can adjust the model view, orient the view, hide and show entities, create and use advanced views, and set various model display options.

**Figure 2.10(a)** View tab

The following list includes some of the View operations you can perform:

*   Orient the model view in the following ways, spin, pan, and zoom models and drawings, display the default orientation, revert to the previously displayed orientation, change the position or size of the model view, change the orientation (including changing the view angle in a drawing), and create new orientations
*   Toggle display style; shading with edges, shading with reflections, shading, hidden line, no hidden line, and wireframe.
*   Highlight items in the graphics window when you select them in the Model Tree
*   Explode or un-explode an assembly view
*   Repaint the Creo Parametric 2.0 graphics window
*   Refit the model to the Creo Parametric 2.0 window after zooming in or out on the model
*   Update drawings of model geometry
*   Hide and unhide entities, and hide or show items during spin or animation

Click: **View** tab > [icons] *off* > [icons] *off* > Appearance Gallery ▼ > **ref_color1** > [brush icon] > **T-NUT.PRT** > **OK** > Display Style ▼ > Wireframe > Display Style ▼ > No Hidden > Display Style ▼ > Hidden Line > Display Style ▼ > Shading > Display Style ▼ > Shading With Reflections > Display Style ▼ > Shading With Edges [Fig. 2.10(b)]

**Figure 2.10(b)** Display Styles

Click:  *on* >  *on* > 🔍 **Refit** > press and hold the **MMB** and rotate the model [Fig. 2.11(a)] > release the **MMB** > 📇 **Named Views** > **Standard Orientation** [Fig. 2.11(b)]

**Figure 2.11(a)** Rotated View

**Figure 2.11(b)** Standard Orientation

## View Tools

Creo Parametric 2.0 provides various view tools, including:

🔍 **Zoom In** Zoom in on the model, 🔍 **Zoom Out** Zoom out from the model, 🔍 **Refit** Adjust the zoom level to fully display the object on the screen, ✋ **Pan** Navigate from one location to another, 📷 **Reorient** Configure model orientation preferences.

Click: **Appearance Gallery ▾** > move your mouse cursor over **<ref-color1>** > **RMB** > **Edit** > (adjust sliders or Enter values) > **Close** > 🔍 **Zoom In** pick two positions about an area you wish to enlarge [Fig. 2.12(a)] > 🔍 **Zoom Out** > 🔍 **Refit** [Fig. 2.12(b)]

**Figure 2.12(a)** Zoom In

**Figure 2.12(b)** Refit

## Using Mouse Buttons to Manipulate the Model

You can also dynamically reorient the model using the **MMB** by itself (**Spin**) or in conjunction with the **Shift** key (**Pan**) or **Ctrl** key (**Zoom, Turn**).

Click:  *off* > hold down **Ctrl** key and **MMB** in the Graphics Window near the model and move the cursor up (zoom out) [Fig. 2.13(a)] > release the **Ctrl** key and **MMB** > hold down **Shift** key and **MMB** in the Graphics Window near the model and move the cursor about the screen (pan) [Fig. 2.13(b)] > release the **Shift** key and **MMB** > hold down **MMB** in the Graphics Window near the model and move the cursor around (spin) [Figs. 2.13(c-d)] > release the **MMB** > press **Ctrl+D** Standard Orientation > *on*

**Figure 2.13(a)** Zoom

**Figure 2.13(b)** Pan

**Figure 2.13(c)** Spin

**Figure 2.13(d)** Spin Again

# System Display Settings

You can make a number of changes to the default colors furnished by Creo Parametric 2.0, customizing them for your own use:

- Define, save, and open color schemes
- Customize colors used in the user interface
- Redefine basic colors used in models
- Assign colors to be used by an entity
- Store a color scheme so you can reuse it
- Open a previously used color scheme

Click: **File > Options** [Fig. 2.14(a)] **> System Colors >** Color scheme: ▾ **> Dark Background >** ▾ **>**
**Black on White >** ▾ **> Default** [Fig. 2.14(b)] **>** ▶ **Graphics** (to expand) **>** ▾ **Graphics** (to collapse)
**>** ▶ **Graphics >** ▶ **Datum >** ▶ **Geometry >** ▶ **Sketcher >** ▶ **Simple Search >** try variations
of different Colors options **>** Color scheme: ▾ **> Default > OK**

Figure 2.14(a) Options                                    Figure 2.14(b) Expanded Options

88

## Information Tools

At any time during the design process you can request model, feature, or other information. Picking on a feature in the graphics window or Model Tree and then **RMB > Info** will provide information about that feature in the Browser. This can also be accomplished by clicking on the feature name in the Model Tree and then **RMB > Info**. Both feature and model information can be obtained using this method. A variety of information can also be extracted using the **Tools** tab Ribbon.

Click: **Model** tab > in the Model Tree, pick once on the chamfer > **RMB > Info > Feature** [Figs. 2.15(a-b)] > ☒ upper right corner of Browser or ⌀ lower left of screen to close Browser > **File > Save > OK** > **File > Manage File > Delete Old Versions >** ✓ > **Ctrl+D**

**Figure 2.15(a)** Info Feature from the Model Tree

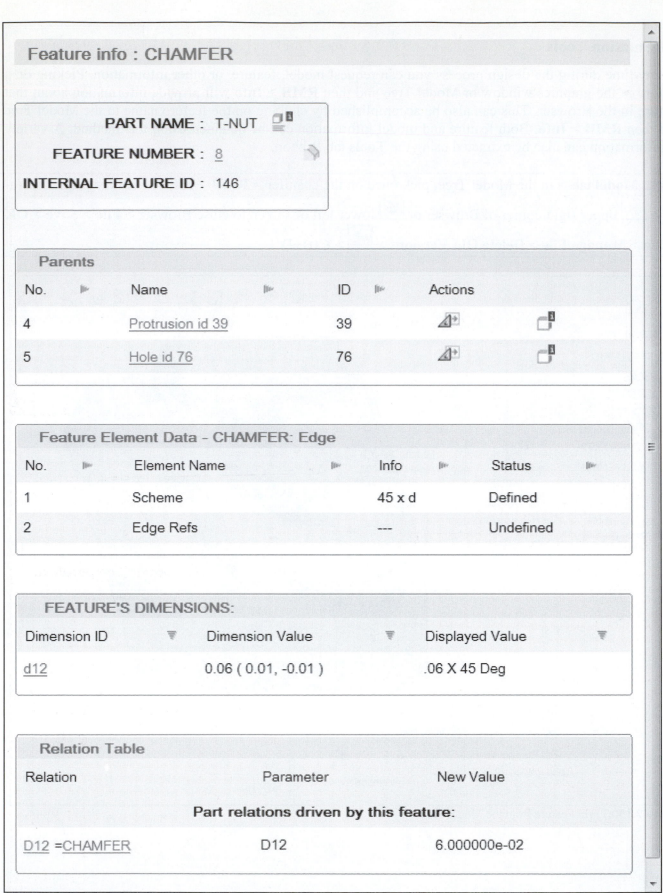

**Feature info : CHAMFER**

PART NAME : T-NUT

FEATURE NUMBER : 8

INTERNAL FEATURE ID : 146

**Parents**

| No. | Name | ID | Actions |
|-----|------|-----|---------|
| 4 | Protrusion id 39 | 39 | |
| 5 | Hole id 76 | 76 | |

**Feature Element Data – CHAMFER: Edge**

| No. | Element Name | Info | Status |
|-----|--------------|------|--------|
| 1 | Scheme | 45 x d | Defined |
| 2 | Edge Refs | --- | Undefined |

**FEATURE'S DIMENSIONS:**

| Dimension ID | Dimension Value | Displayed Value |
|--------------|-----------------|-----------------|
| d12 | 0.06 ( 0.01, -0.01 ) | .06 X 45 Deg |

**Relation Table**

| Relation | Parameter | New Value |
|----------|-----------|-----------|
| Part relations driven by this feature: | | |
| D12 =CHAMFER | D12 | 6.000000e-02 |

**Figure 2.15(b)** Feature Info: Chamfer, Displayed in the Browser

In the Model Tree, select: **T-NUT.PRT > RMB > Info > Model** [Figs. 2.15(c-d)] > ☒ close Browser

**Figure 2.15(c)** Model Information

**Model Info : T-NUT**

| PART NAME : | | T-NUT | | | | ☐ | |

| Units: | | Length: | Mass: | Force: | | Time: | Temperature: |
|---|---|---|---|---|---|---|---|
| Inch lbm Second (Creo Parametric Default) | | in | lbm | in lbm / sec^2 | | sec | F |

**Feature List**

| No. ▶ | ID ▶ | Name ▶ | Type ▶ | | Actions | | Sup Order | Status |
|---|---|---|---|---|---|---|---|---|
| 1 | 33 | DTM1 | DATUM PLANE | | ◢⁺ | ⊟ | --- | Regenerated |
| 2 | 35 | DTM2 | DATUM PLANE | | ◢⁺ | ⊟ | --- | Regenerated |
| 3 | 37 | DTM3 | DATUM PLANE | | ◢⁺ | ⊟ | --- | Regenerated |
| 4 | 39 | — | PROTRUSION | | ◢⁺ | ⊟ | --- | Regenerated |
| 5 | 76 | — | HOLE | | ◢⁺ | ⊟ | --- | Regenerated |
| 6 | 96 | — | CHAMFER | | ◢⁺ | ⊟ | --- | Regenerated |

**Figure 2.15(d)** Information from the Model

## The Model Tree

The **Model Tree** is a tabbed feature on the Creo Parametric 2.0 Navigator that displays a list of every feature or part in the current part, assembly, or drawing.

    The model structure is displayed in hierarchical (tree) format with the root object (the current part or assembly) at the top of its tree and the subordinate objects (parts or features) below. If you have multiple Creo Parametric 2.0 windows open, the Model Tree contents reflect the file in the current active window. The Model Tree lists only the related feature and part level items in a current object and does not list the entities (such as edges, surfaces, curves, and so forth) that comprise the features. Each Model Tree item contains an icon that reflects its object type, for example, hidden, assembly, part, feature, or datum plane (also a feature). The icon can also show the display status for a feature, part, or assembly, for example, suppressed.

    Selection in the Model Tree is object-action oriented; you select objects in the Model Tree without first specifying what you intend to do with them. You can select components, parts, or features using the Model Tree. You cannot select the individual geometry that makes up a feature (entities). To select an entity, you must select it in the graphics window.

    With the **Settings** tab you can control what is displayed in the Model Tree. You can add informational columns to the Model Tree window, such as **Tree Columns** containing parameters and values, assigned layers, or feature name for each item. You can use the cells in the columns to perform context-sensitive edits and deletions. These options will be covered elsewhere in the text, as they are needed in the design process.

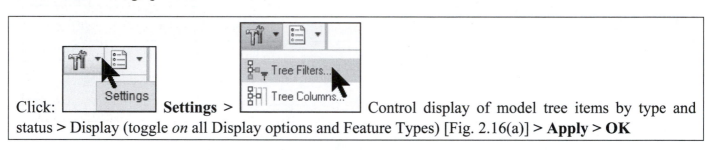

Click: [Settings] **Settings >** [Tree Filters... / Tree Columns...] Control display of model tree items by type and status > Display (toggle *on* all Display options and Feature Types) [Fig. 2.16(a)] > **Apply > OK**

**Figure 2.16(a)** Model Tree Items Dialog Box

Click:  **Show** > [Expand All] > **Show** > [✓ Show PopUp Viewer] > in the Model Tree, select **T-NUT.PRT** > move your mouse cursor on top of the icon to the left of **T-NUT.PRT** to see the pop-up view [Fig. 2.16(b)] > [icon] > [Tree Columns...] [Fig. 2.16(c)]

**Figure 2.16(b)** Pop-up Viewer

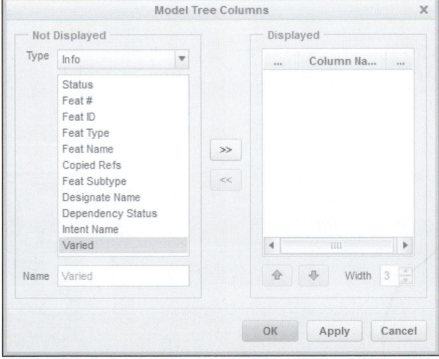

**Figure 2.16(c)** Model Tree Columns Dialog Box

93

Click: **Feat #** > [ >> ] > **Feat ID** > [ >> ] > **Feat Type** > **Enter** > **Feat Name** > **Enter** [Fig. 2.16(d)] > **OK**

| | Feat # |
|---|---|
| ☐ T-NUT.PRT | ↔ |
| ▱ DTM1 | 1 |

> press **LMB** and drag the Model Tree Panel to see all the Columns [Fig. 2.16(e)] > press **LMB** and pull the panel back until only the Model Tree and **Feat #** is displayed

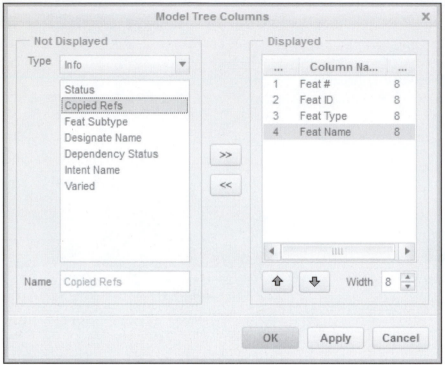

**Figure 2.16(d)** Adding Displayed Columns

| Model Tree | Feat # | Feat ID | Feat Type | Feat Name | |
|---|---|---|---|---|---|
| ☐ T-NUT.PRT | | | | | |
| ▱ DTM1 | 1 | 33 | Datum Plane | DTM1 | ↔ |
| ▱ DTM2 | 2 | 35 | Datum Plane | DTM2 | |
| ▱ DTM3 | 3 | 37 | Datum Plane | DTM3 | |
| ▼ ⬠ Protrusion id 39 | 4 | 39 | Protrusion | | |
| ✎ Section 1 | <None> | | | | |
| ⬚ Hole id 76 | 5 | 76 | Hole | | |
| ⬎ Chamfer id 96 | 6 | 96 | Chamfer | | |
| ⬎ Round id 127 | 7 | 127 | Round | | |
| ⬎ Chamfer id 146 | 8 | 146 | Chamfer | | |
| ➜ Insert Here | | | | | |
| ▼ ⬡ Footer | <None> | | | | |
| ▼ ⬡ INTERFACES | <None> | | | | |
| ▼ ⬡ INTFC001 | 9 | 166 | Component I... | INTFC001 | |
| ⊥ Coincident | | | | | |
| ⊥ Coincident | | | | | |

**Figure 2.16(e)** Model Tree and Model Tree Columns

# Working on the Model

In Creo Parametric 2.0, you can select objects to work on from within the graphics window or in the Model Tree by using the mouse or the keyboard. The object types that are available for selection vary depending on whether you select an object from within the graphics window or in the Model Tree. You can select any type of object, including features, 3-D notes (model Annotations), parts, datum objects (planes, axes, curves, points, and coordinate systems), and geometry (edges and surfaces) from within the graphics window. Additionally, since the Model Tree displays only parts, components, and features, you can select only those object types from within the Model Tree.

Selection in both the graphics window and the Model Tree can be action-object or object-action oriented, depending on the process you choose within Creo Parametric 2.0 to build your model. You can specify the action you want to perform on an object before you select the object, or you can select the object before you specify the action.

You can *dynamically* modify features from within the graphics window as you work. Features can be edited by selecting them from the graphics window directly, or from the Model Tree.

---

Click: 🔍 **Refit** > pick on the hole in the Graphics Window [Fig. 2.17(a)] > press and hold **RMB** anywhere in the Graphics Window > **Delete** > **OK** in the Delete dialog box [Fig. 2.17(b)] > **Delete Rels** [Fig. 2.17(c)] > ↺ **Undo** from the Quick Access Toolbar [Fig. 2.17(d)]

---

**Figure 2.17(a)** Select the Hole

**Figure 2.17(b)** RMB > Delete

**Figure 2.17(c)** Deleted Hole

**Figure 2.17(d)** Undo, Hole Returns

Click on the protrusion in the Model Tree > press **RMB** > **Edit Definition** [Fig. 2.18(a)] > press and hold the **LMB** on a white nodal handle and drag to change the protrusion's length from **1.125** [Fig. 2.18(b)] to **1.50** [Fig. 2.18(c)] > release the **LMB** >  > anywhere in the Graphics Window, **LMB** to deselect [Fig. 2.18(d)] > **Ctrl+D** > **Ctrl+S** > **OK**

**Figure 2.18(a)** Redefining a Feature

**Figure 2.18(b)** Dynamically Redefining a Feature

**Figure 2.18(c)** Dynamically Dragging a Nodal Handle

**Figure 2.18(d)** Protrusion Redefined

Click: 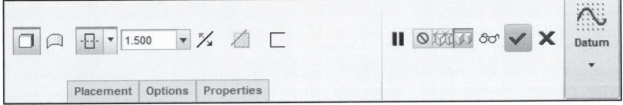 > **RMB** > **Edit Definition** > **Options** tab [Fig. 2.19(a)] > change **1.50** to **1.25** [Fig. 2.19(b)] > **Enter** > ☑ > **LMB** anywhere in Graphics Window to deselect [Fig. 2.19(c)] > double-click on the model [Fig. 2.19(d)] (you can double-click on any dimension and change its value) > **LMB** anywhere in Graphics Window > **File** > **Save** > **OK**

**Figure 2.19(a)** Extrude Dashboard

**Figure 2.19(b)** Options Tab

**Figure 2.19(c)** Edited Model

**Figure 2.19(d)** Editing Model Dimensions

# Flexible Modeling

Flexible modeling allows you to experiment with your design without being committed to changes. You can make explicit modifications to selected geometry while ignoring pre-existing relationships. Use the buttons in the following groups to quickly select and manipulate geometry.

- **Recognize and Select**—selection only of the specified type of geometry.
- **Transform**—direct manipulation of the selected geometry.
- **Recognition**—recognizing patterns and symmetry so that when one member is modified, the modification is propagated to all pattern members or symmetrical geometry.
- **Edit Features**—editing of the selected geometry or surface.

The intention is that when you enter a tool, you can perform the required tasks using the right-click shortcut menus. The procedures for each tool use this approach since it is the fastest way to work and is sufficient for most tasks. Use the tabs to select rule–based surface sets, to view selections and references, and to change the default system selections. Detailed information about the individual tool tabs can be found in the About the User Interface topics for each tool.

*If you have the Creo Parametric 2.0 Flexible License activated on your system, the following commands will be available.*

---

Click: ▦ to close the Model Tree *(lower left-hand corner)* > **Flexible Modeling** tab [Fig. 2.20(a)] >

**Ctrl+D** > ⬛ **Move using Dragger** > select the front face [Fig. 2.20(b)] > place your mouse cursor on the *green* circle, hold down the **LMB** and drag the face to rotate [Fig. 2.20(c)] > release the **LMB**

---

**Figure 2.20(a)** Flexible Modeling Tab

**Figure 2.20(b)** Select the Front Face

**Figure 2.20(c)** Rotate the Face with Green Circle

Drag the face by rotating the *red* circle [Fig. 2.20(d)] > pull the face with the *blue* Ball [Fig. 2.20(e)] > spend some time and experiment with this tool > ✅ [Fig. 2.20(f)] > **Ctrl+S** > **OK** > **File** > **Save As** > Type ▾ > **Zip File (\*.zip)** > **OK** > upload your version of the part to your course interface or attach to an email and send to your instructor and/or yourself > 🗁 **Close** > 🔲 to open the Navigator/Model Tree *(lower left-hand corner)*

**Figure 2.20(d)** Rotate the Face with Red Circle

**Figure 2.20(e)** Pull the front Face with the Blue Ball

**Figure 2.20(f)** Flex Move Completed

## Productivity Enhancements

Customizing your Creo Parametric 2.0 screen interface and setting your mouse cursor to go to the default position are two productivity enhancements that can be employed to get your design created and documented quickly.

As far as input devices, the mouse is still a mouse, but- I encourage my students and trainees to buy a 5-Button mouse. Of course, many of you reading this already have a fancy Space Ball, etc. These input devices greatly improve your productivity as well as cut down on the amount of mouse picks, etc. But, for those with limited resources, the 5-button mouse for about 25.00 to 50.00 dollars (US) will be all that is needed to fly with Creo Parametric 2.0. Program your 4th and 5th button to initiate commands that you frequently use- Delete, Back, Undo, Recall Window, or whatever will reduce your need to move your mouse for the next input. I use the Back and Recall Window most since I want to quickly go back to previous Browser screens and also I flip programs when writing. Production oriented drafters-designers-engineers, will have different needs and thus different button assignments.

A simple way to eliminate much of your mouse movement by your hand and arm is to set the mouse cursor to always go to the default position after a pick. Every command pick will move the pointer to the default option on the next menu.

From your computer Desktop, click: **Start > Control Panel > Mouse > Pointer Options >** Snap To

(Fig. 2.21) > **OK** > ⊠ close the Control Panel

**Figure 2.21** Mouse Properties *(the appearance of this dialog box may be different depending on your operating system)*

100

# Customizing the User Interface (UI)

Customize your user interface to increase your efficiency in modeling. You can customize the Creo Parametric 2.0 user interface, according to your needs or the needs of your group or company, to include the following:

- Create keyboard macros, called *mapkeys*, and add them to the menus and toolbars
- Add or remove existing toolbars
- Move or remove commands from the menus or toolbars
- Add options to the Menu Manager
- Blank (make unavailable) options in the Menu Manager
- Set default command choices for Menu Manager menus

Press: **Ctrl+O** > **t-nut.prt** > **Open** > **View** tab > [icons] *off* > move pointer anywhere in the Graphics Toolbar > **RMB** [Fig. 2.22(a)] > all options *on* [Fig. 2.22(b)] > **LMB** in Graphics Window > [icon] > [icon] > Shading With Edges > [icons] datum features/spin center *on*

**Figure 2.22(a)** Customize Graphics Toolbar *(the order may appear differently for you)*     **Figure 2.22(b)** All Options On

To change the location of the Graphics Toolbar, click: **RMB** within the Graphics Toolbar > **Location** [Fig. 2.22(c)] > **Size** [Fig. 2.22(d)] > **Large** [Fig. 2.22(e)] > **RMB** within the Graphics Toolbar > **Reset to Default** [Fig. 2.22(f)]

**Figure 2.22(c)** Graphics Toolbar Location

**Figure 2.22(d)** Graphics Toolbar Size

**Figure 2.22(e)** Graphics Toolbar Size Large *(the order may appear differently for you)*

**Figure 2.22(f)** Graphics Toolbar Reset to Default *(the order may appear differently for you)*

Place your pointer anywhere inside the **View** tab in the Ribbon [Fig. 2.23(a)] > **RMB** > check **Hide**

**Command Labels** > > place your pointer anywhere inside the Ribbon > **RMB** > check **Minimize the Ribbon** [Fig. 2.23(b)] > place your pointer anywhere inside the Ribbon tabs > **RMB** > uncheck **Minimize the Ribbon** > place your pointer anywhere inside the **View** tab > **RMB** > uncheck **Hide Command Labels** > place your pointer anywhere inside the Ribbon tabs > **RMB** > **Tabs** [Fig. 2.23(c)] leave all Tabs active > **LMB** anywhere in the Graphics Window

**Figure 2.23(a)** Customize the Ribbon via the View tab

**Figure 2.23(b)** Minimized Ribbon

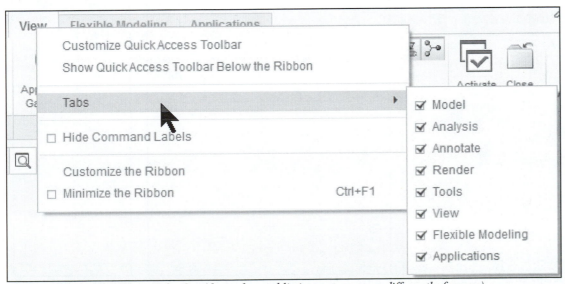

**Figure 2.23(c)** All Ribbon Tabs On *(the order and listing may appear differently for you)*

Click: **Model** tab > **Datum** Group *(Groups that have a* ⏷ *indicate that there are overflow commands for that Group- more commands which are not typically placed on the Ribbon)* > **RMB** > **Groups** [Fig. 2.24(a)] > **Customize the Ribbon** > move the Creo Parametric Options dialog box down > **Datum** Group [Fig. 2.24(b)] > pick on **Curve** and drag and drop [Fig. 2.24(c)] into the Ribbon [Fig. 2.24(d)] > **OK** to close the Creo Parametric Options dialog box

*(After you are familiar with Creo Parametric 2.0's default UI, you can customize the interface based on your or your companies' productivity requirements.)*

**Figure 2.24(a)** Groups

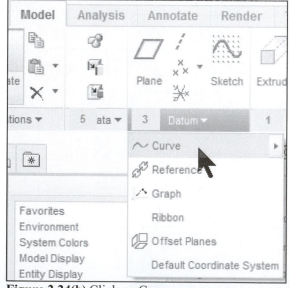

**Figure 2.24(b)** Click on Curve

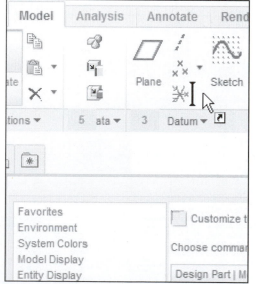

**Figure 2.24(c)** Drag and Drop Curve

**Figure 2.24(d)** Curve

Click: **File > Options > Customize Ribbon > New Tab >** select **New Tab (Custom) > RMB > Rename > *type* Production** [Fig. 2.25(a)] **> OK >** select **New Group (Custom) > RMB > Rename > *type* Creo**

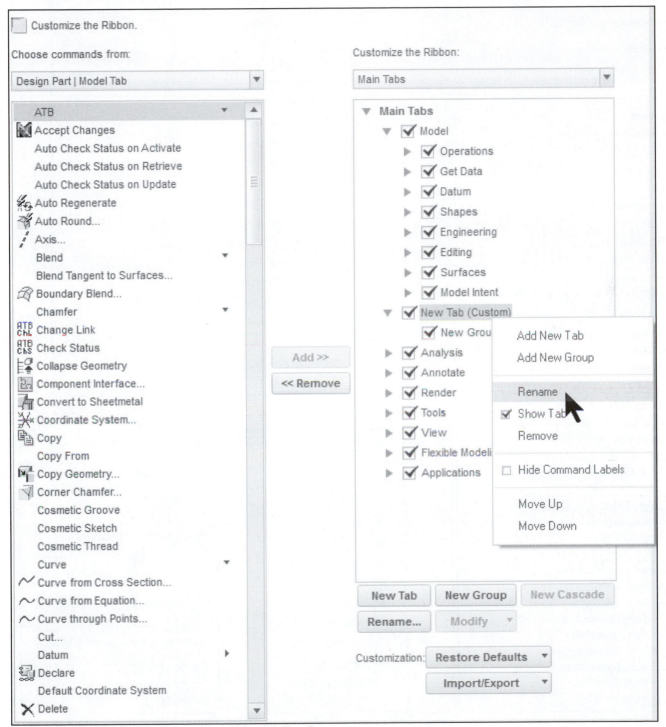

**Figure 2.25(a)** Customize Ribbon, New Tab > RMB > Rename

Click: **Edit Definition** > **Add** [Fig. 2.25(b)] > Choose commands from: **Design Part | Model Tab** > **All Commands** [Fig. 2.25(c)] > **Delete Old Versions** > **Add** [Fig. 2.25(d)] > **OK** > **Production** tab > **Delete Old Versions** [Fig. 2.25(e)] > **Enter** [Fig. 2.25(f)] *(customize your Production tab after you see what commands you would like faster access to)*

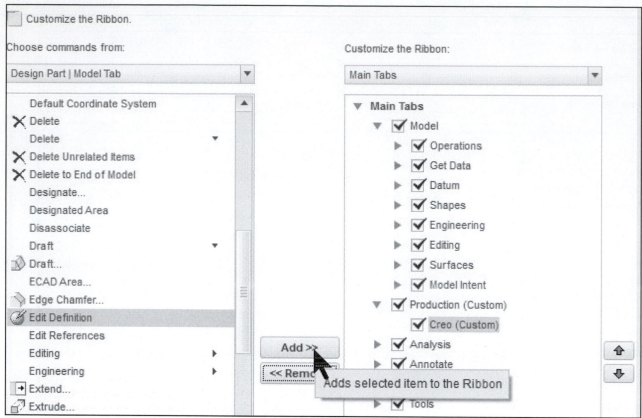

**Figure 2.25(b)** Edit Definition > Add

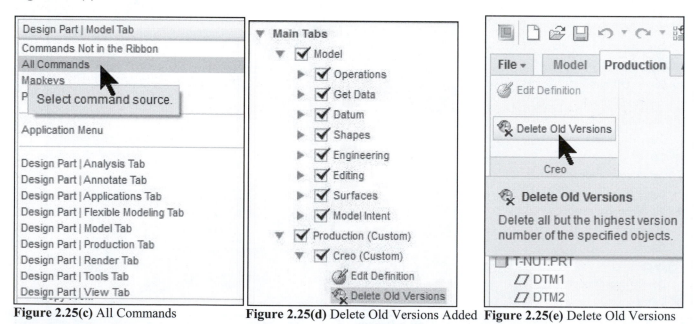

**Figure 2.25(c)** All Commands   **Figure 2.25(d)** Delete Old Versions Added   **Figure 2.25(e)** Delete Old Versions

**Figure 2.25(f)** Delete Old Versions Confirmation

# Mapkeys

A **Mapkey** is a macro that maps frequently used command sequences to certain keyboard keys or sets of keys. Mapkeys are saved in the configuration file, and are identified with the option *mapkey*, followed by the identifier and then the macro. You can define a unique key or combination of keys which, when pressed, executes the mapkey macro. You can create a mapkey for virtually any task you perform frequently within Creo Parametric 2.0. By adding mapkeys to your Ribbon or Toolbars, you can run mapkeys with a single mouse click or menu command and thus streamline your workflow.

Click: **File > Options > Environment > Mapkeys Settings** [Fig. 2.26(a)] > **New** from the Mapkeys dialog box [Fig. 2.26(b)] (Record Mapkey dialog box opens) > Key Sequence, *type* **$F6** > Name, *type* **Display** > Description, *type* **Change display** [Fig. 2.26(c)] > **Record**

**Figure 2.26(a)** Creo Parametric Option Dialog Box > Environment > Mapkeys Settings

Click: **View** tab >  toggle *off* > **File** > **Options** > **Model Display** > **Trimetric** > **Isometric** > **Remove hidden lines** > **OK** > **No** > **Stop** (from Record Mapkey dialog box) > **OK** (from Record Mapkey dialog box) > Save Mapkeys **Changed** (from Mapkeys dialog box) > File name: *type* **CREO_textbook.pro** > **OK** > **Close** (from Mapkeys dialog box) > press **F6** > press **F6** [Fig. 2.26(d)]

Figure 2.26(b) Mapkeys          Figure 2.26(c) Record Mapkey          Figure 2.26(d) New Display

Click: **File** > **Options** > **Customize Ribbon** > **Design Part | Model Tab** >

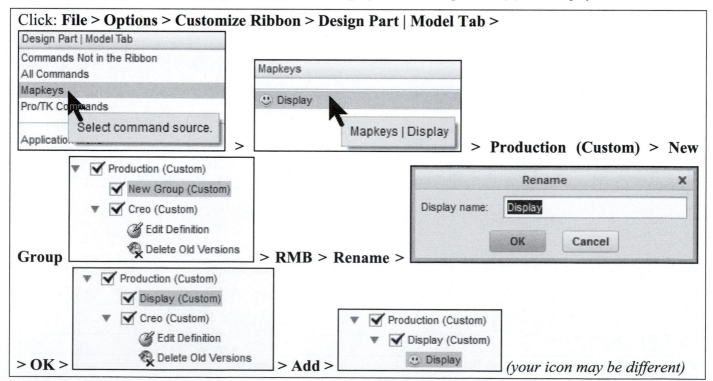

> **Production (Custom)** > **New**

**Group** ... > **RMB** > **Rename** >

> **OK** > ... > **Add** > ... *(your icon may be different)*

Click: **Modify > Choose Button Image** [Fig. 2.26(e)] > select the red diamond ( )

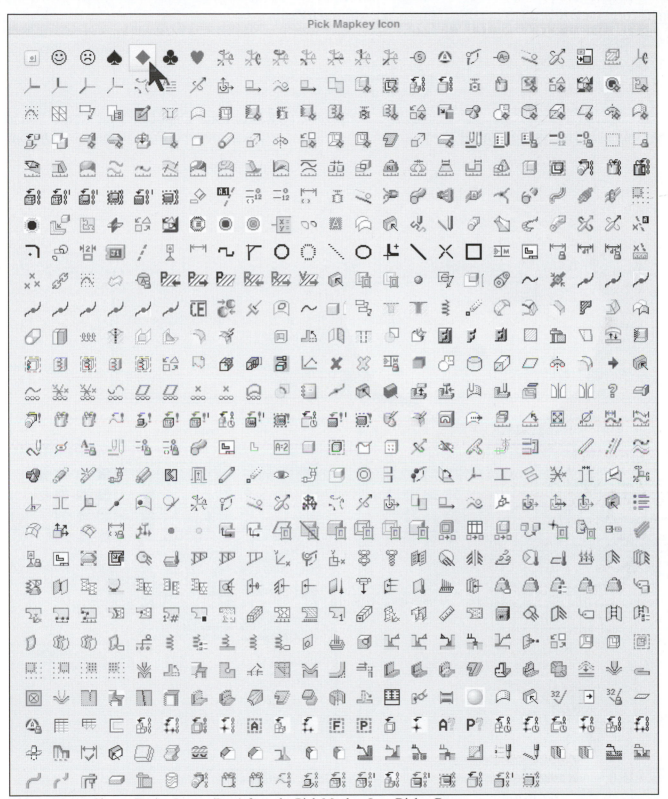

**Figure 2.26(e)** Choose Button Image (Icon) from the Pick Mapkey Icon Dialog Box

Click: **Modify > Edit Button Image** [Fig. 2.26(f)] > click on a color block and edit the picture as desired

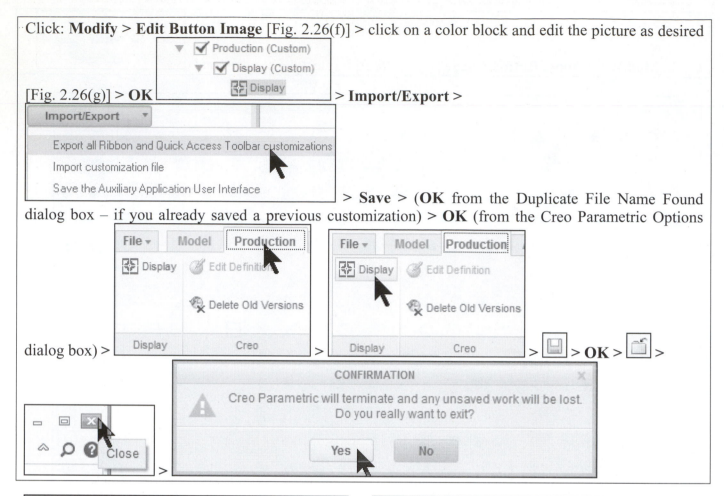

[Fig. 2.26(g)] > **OK** > **Import/Export** >

> **Save** > (**OK** from the Duplicate File Name Found dialog box – if you already saved a previous customization) > **OK** (from the Creo Parametric Options dialog box) > > **OK** >

>

**Figure 2.26(f)** Button Editor

**Figure 2.26(g)** Edit the Picture

Download another model from the PTC Catalog or *www.cad-resources.com* and practice navigating the interface and using commands introduced in this Lesson.

# Lesson 3 Direct Modeling

**Figure 3.1** Ten Quick Modeling Parts

## OBJECTIVES

- Modeling simple parts quickly using **default selections**
- Sample **Feature Tools**
- Try out a variety of **Engineering Tools** including: **Hole**, **Shell**, **Round**, **Chamfer**, and **Draft**
- Sketch simple **sections**

## REFERENCES AND RESOURCES

For **Resources** go to **www.cad-resources.com** > click on the Creo Parametric 2.0 Book cover

- Lesson 3 Lecture
- Lesson 3 3D models embedded in a PDF
- Creo Parametric 2.0 Quick Reference Card
- Creo Parametric 2.0 Configuration Options

## Modeling

The purpose of this lesson is to quickly introduce you to a variety of **Feature Tools** and **Engineering Tools**. You will model a variety of very simple parts (Fig. 3.1). Little or no explanation of the methodology or theory of the Tool or process will accompany the instructions. By using mainly default selections; you will create models that will display the power and capability of Creo Parametric 2.0.

In the next lesson, a detailed step-by-step systematic description will accompany all commands. Here, we hope to get you up and running on Creo Parametric 2.0 without any belabored explanations.

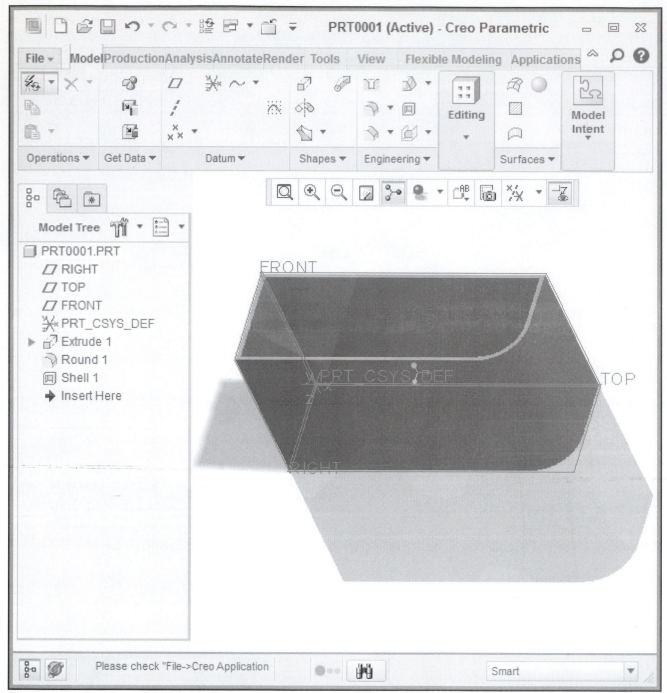

**Figure 3.2(a)** Part One – PRT0001.PRT

## Part Model One (PRT0001.PRT) (Extrude)

This model [Fig. 3.2(a)] will introduce the **Extrude Tool** to create a simple box-shape, the **Round Tool** to add a round to one edge, and the **Shell Tool** to remove one surface and make the part walls a consistent thickness. In this direct modeling example, we will provide step-by-step illustrations. For subsequent parts, only important steps or illustrations displaying aspects of the command sequence that represent new material will be provided. The same applies to *Tool Tips* that appear as you pass the mouse pointer (cursor) over a button or icon. All parts have been created with out-of-the-box system settings for Creo Parametric 2.0 including default templates, grid settings, and so forth.

Launch **Creo Parametric 2.0** >  **Select Working Directory** > if needed, select the working directory > **OK** > New Create a new model > (use the default name **prt0001**) > > **OK** > **File** > **Options** > **Model Display** > **Show datums** checked *on* > Set shade quality to: **50** (note- make this 5 to keep the size of your files smaller) > **OK** > **No** > **View** tab > *on* > *on* > select datum **FRONT** in the Model Tree [Fig. 3.2(b)]

*(Do not change directories if you have one assigned as a default by your school or company.)*

**Figure 3.2(b)** Pre-select the FRONT Datum Plane in the Model Tree

Click: **Model** tab > Extrude the pre-selected FRONT datum is the Sketch Plane, the sketch view remains in a 3D trimetric orientation > in the Graphics Window, press **RMB** > **Corner Rectangle** [Fig. 3.2(c)]

**Figure 3.2(c)** RMB > Corner Rectangle

Pick the first corner of the rectangle [Fig. 3.2(d)] > pick the second corner of the rectangle [Fig. 3.2(e)]

**Figure 3.2(d)** First Corner of the Rectangle at the origin

**Figure 3.2(e)** Second Corner of the Rectangle

Click: **MMB** to deselect the Corner Rectangle command > press **RMB** > **OK** [Fig. 3.2(f)] *(if your feature looks too thin, move a drag handle: place the pointer on the square drag handle > press and hold down the LMB > move the square drag handle forward to give the feature greater depth > release the LMB) >*

**Display Style** from the Graphics Toolbar > Shading [Fig. 3.2(g)] >

**Figure 3.2(f)** Shaded (Closed Loop) Rectangular Section

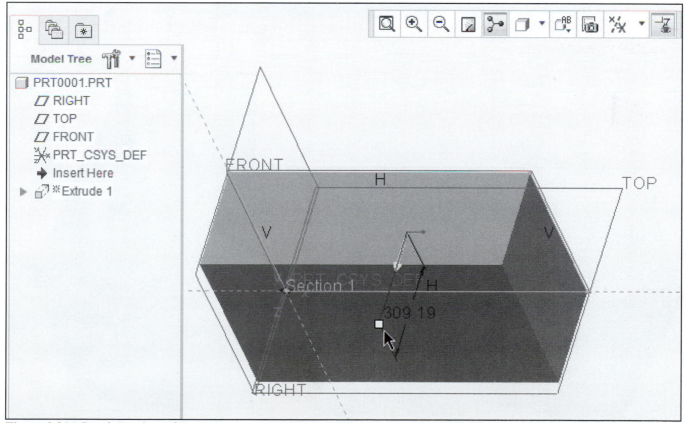

**Figure 3.2(g)** Depth Preview of Extrusion

With the extrusion *still selected*, pick on the right front edge of the part [Fig. 3.2(h)] > press **RMB** > **Round Edges** [Fig. 3.2(i)] > move a drag handle [Fig. 3.2(j)] > enlarge the round [Fig. 3.2(k)] > **MMB**

**Figure 3.2(h)** Pick on the Front Edge

**Figure 3.2(i)** Round Edges

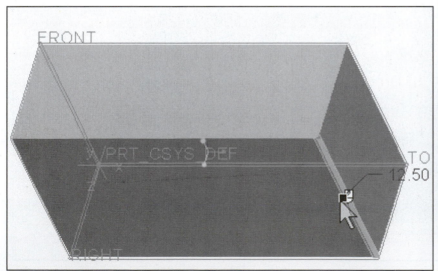

**Figure 3.2(j)** Move the Radius Drag Handle

**Figure 3.2(k)** Enlarged Round

116

With the round feature *still selected*, pick on the top surface > slightly move the mouse pointer and pick again (the surface will highlight) [Fig. 3.2(l)] > Shell [Fig. 3.2(m)] > **MMB** [Fig. 3.2(n)] > **Ctrl+S** > **OK** > **File** > **Save As** > Type ▼ > **Zip File (\*.zip)** > **OK** > **File** > **Close**

**Figure 3.2(l)** Selected Top Surface is Highlighted

**Figure 3.2(m)** Shell Tool Applied

**Figure 3.2(n)** Completed Part

## Part Model Two (PRT0002.PRT) (Draft)

Press: **Ctrl+N** > **OK** (prt0002) > pick on the **TOP** datum from the Graphics Window >  Extrude > in the Graphics Window, press **RMB** > **Circle** [Fig. 3.3(a)] > press **RMB** (note Circle is checked *on*) [Fig. 3.3(b)] > release the **RMB** > slightly move the mouse pointer in the Graphics Window > **RMB**

**Figure 3.3(a)** RMB > Circle          **Figure 3.3(b)** Circle checked on

Sketch a circle by picking at the origin [Fig. 3.3(c)] and then stretching the circle's diameter to a convenient size [Fig. 3.3(d)] > pick again to establish the circle's diameter > slightly move the mouse pointer in the Graphics Window > **MMB** to deselect the Circle command [Fig. 3.3(e)]

**Figure 3.3(c)** Pick the Center of the Circle          **Figure 3.3(d)** Pick to Establish the Circle's Diameter

Click: [Fig. 3.3(f)] > **Options** tab > **Add Taper** checked *on* > click in the taper value field > *type* **3** >

Enter [Fig. 3.3(g)] > **Verify** *on* > **Verify** *off* > **Options** tab > **Add Taper** checked *off* > > > in the Graphics Window, **LMB** to deselect

*(Notes: There are angle limitations to adding a taper versus creating a draft.)*

**Figure 3.3(e)** Circle Sketch          **Figure 3.3(f)** Depth Preview of Extruded Circle

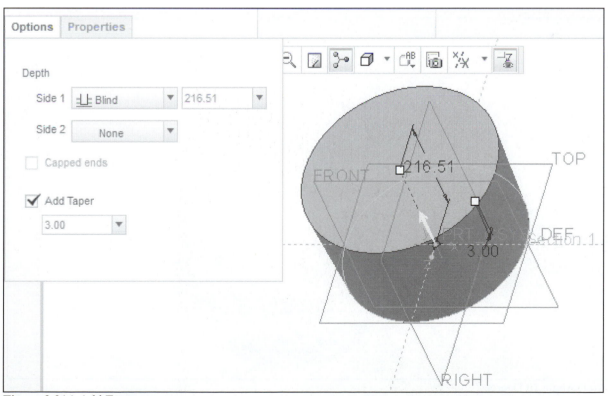

**Figure 3.3(g)** Add Taper

119

Click: 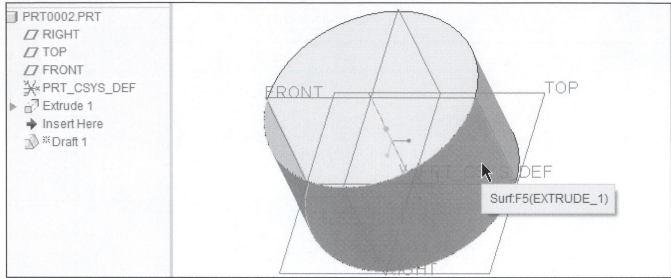 Draft > pick the front surface of the cylinder [Fig. 3.3(h)] > press **RMB** > **Draft Hinges** > pick the **TOP** datum as the Draft Hinge > double-click on the **1.0** degree dimension in the Graphics Window or click in the Angle value field > *type:* **10** [Fig. 3.3(i)] > **Enter** > ☑ > **Ctrl+S** > **Enter**

*(Notes: The default shade color has been lightened for illustration capture and printing purposes.)*

**Figure 3.3(h)** Vertical Surface of the Cylindrical Part Selected as the Draft Surface

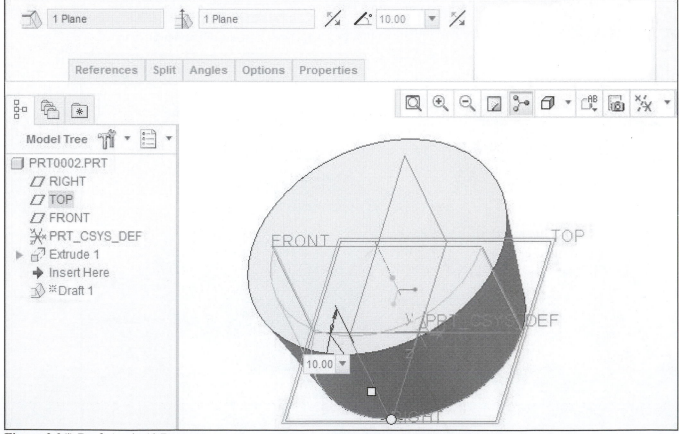

**Figure 3.3(i)** Draft Angle **10** Degrees

With the Draft feature *still selected*, pick on the top edge [Fig. 3.3(j)] > 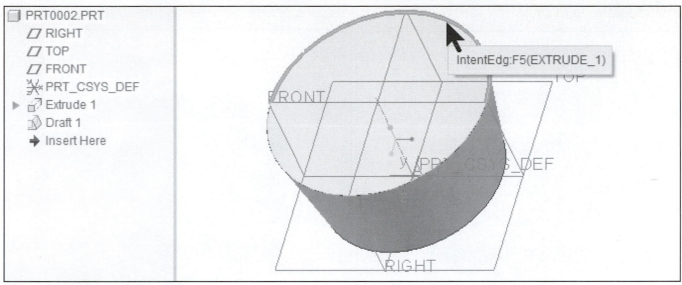 Chamfer > click in the D value field > *type* **50** > **Enter** [Fig. 3.3(k)] > ✓ > **Ctrl+S** > **Enter**

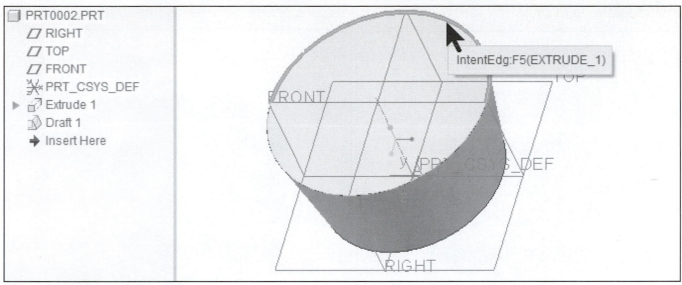

**Figure 3.3(j)** Pick on the Top Edge

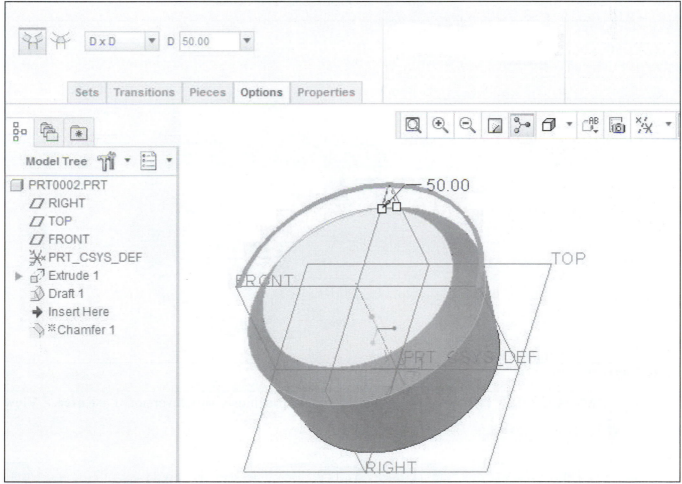

**Figure 3.3(k)** Chamfer Preview

Press and hold the **MMB** and spin the part to display the bottom surface [Fig. 3.3(l)] > release the **MMB** > pick on the bottom surface > slightly move the mouse pointer and pick the same surface again > 
Shell  [Fig. 3.3(m)] >  [Fig. 3.3(n)] > **Ctrl+D** > in the Graphics Window, **LMB** to deselect > **Ctrl+S** > **OK**

**Figure 3.3(l)** Spin the Model

**Figure 3.3(m)** Previewed Shell

**Figure 3.3(n)** Completed Shell

Click: **File > Manage File > Delete Old Versions** (removes previously saved versions) > **Enter > View** tab > **Display Style ▼** > **Shading With Reflections** > **Display Style ▼** > **Shading With Edges** > **File > Save As >** Type > **Zip File (\*.zip) > OK > File > Close > File > Manage Session > Erase Not Displayed > OK**

# Part Model Three (PRT0003.PRT) (Hole)

Press: **Ctrl+N** (prt0003) > **OK** > preselect the **FRONT** datum > **RMB** in the Ribbon > **Customize the Ribbon** > move the Creo Parametric Options dialog down if needed > **RMB** on the Extrude button in the Shapes group of the Model tab Ribbon > **Small Button** > **OK** (in the Creo Parametric Options dialog box) > [Extrude] from the Ribbon > in the Graphics Window, press **RMB** > **Corner Rectangle** > sketch a rectangle by defining two corners > **MMB** > **LMB** to deselect [Fig. 3.4(a)] > press **RMB** > **Fillet** > pick the two lines that form the upper right-hand corner [Fig. 3.4(b)] > ✓ [Fig. 3.4(c)] > move the drag handle forward to increase the model's depth [Fig. 3.4(d)] > ✓ > **Ctrl+S** > **Enter**

**Figure 3.4(a)** Sketch a Rectangle

**Figure 3.4(b)** Create a Fillet

**Figure 3.4(c)** Depth Preview

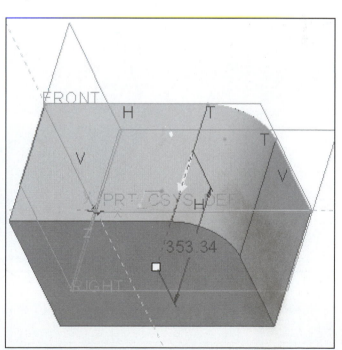

**Figure 3.4(d)** Move the Drag Handle to change the Depth

Click: **View** tab >  **ptc-ceramic** > pick on the part name (**PRT0003.PRT**) in the Model Tree > in the Graphics Window, click: **MMB** > **Model** tab > | Axis | > pick on the cylindrical surface [Fig. 3.4(e)] > **OK** [Fig. 3.4(f)]

**Figure 3.4(e)** Pick on the Cylindrical Surface to Establish the References for a Datum Axis

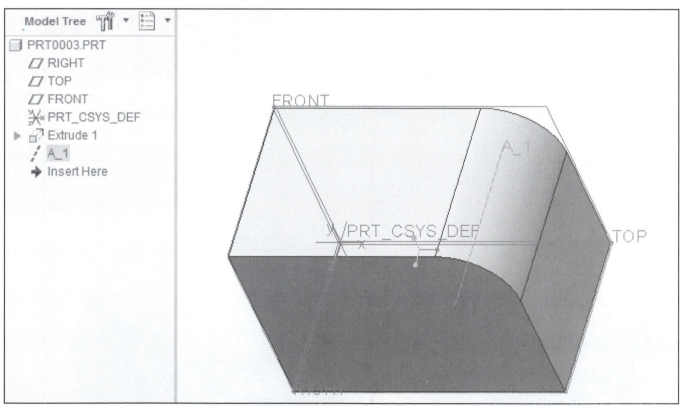

**Figure 3.4(f)** Axis A_1 Created and still Selected

With the datum axis *still highlighted/selected*, click: [ Hole ] [Fig. 3.4(g)] > [ ] expand depth options by opening slide-up panel > **Drill to intersect with all surfaces** > [ Placement ] tab > while holding down the **Ctrl** key, pick on the front *surface (not the FRONT Datum Plane)* [Fig. 3.4(h)] > release the **Ctrl** key

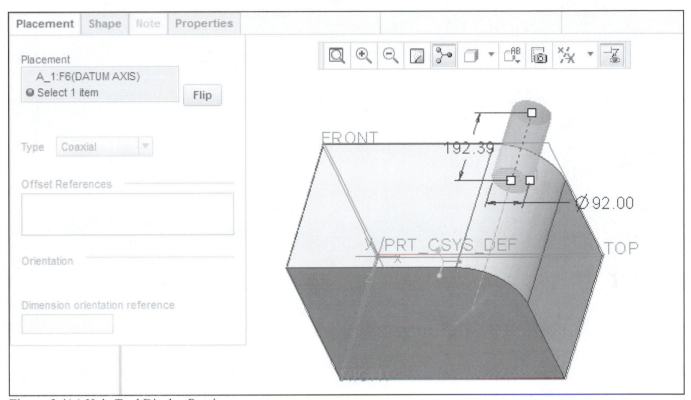

**Figure 3.4(g)** Hole Tool Display Preview

**Figure 3.4(h)** Placement tab, Front surface selected

Click:  > with the hole *still selected*; in the Graphics Window, press: **RMB > Convert to Lightweight** [Figs. 3.4(i-j)] >  **Undo > View** tab > **Display Style > Shading with Reflections** [Fig. 3.4(k)] > **Display Style > Shading with Edges > Ctrl+S > Enter > File > Manage File> Delete Old Versions > Enter > File > Save As > Type**  **> Zip File (*.zip) > OK > File > Close**

**Figure 3.4(i)** Convert the Hole Display to Lightweight *(your Hole number may be different)*

**Figure 3.4(j)** Lightweight Display of Hole. Note the change in the Model Tree Representation of the Hole Symbol

**Figure 3.4(k)** Shading with Reflections (Enhanced Realism) *(the quality of your graphics card and graphics settings may prevent this display)*

# Part Model Four (PRT0004.PRT) (Cut)

Press: **Ctrl+N** (prt0004) > **OK** > **View** tab > **Display Style** > **Shading with Edges** > **Model** tab > **Extrude** > **Placement** from the Dashboard > **Define...** > select the **FRONT** datum from the Model Tree > **Sketch** > **Sketch View** from the Graphics Toolbar > **Line Chain** > starting at the origin, sketch the outline using five lines forming a *closed section* [Fig. 3.5(a)] > **MMB** to end the current line > **MMB** to end the current tool and to display weak dimensions *[section will be shaded if done correctly]* > **LMB** to deselect > **Ctrl+D** [Fig. 3.5(b)] > press **RMB** > **OK** [Fig. 3.5(c)] > **Enter** > **Ctrl+S** > **Enter** > in the Graphics Window, **LMB** to deselect

**Figure 3.5(a)** Sketch the Five Lines of the Enclosed Section *(your weak dimensions will be different)*

**Figure 3.5(b)** Sketch Preview in the Standard Orientation View

**Figure 3.5(c)** Depth Preview

Pick on the top horizontal surface > move the pointer slightly and pick again [Fig. 3.5(d)] > Extrude > > Hidden Line > > in the Graphics Window, press **RMB** > **Corner Rectangle** > sketch a rectangle > **MMB** [Fig. 3.5(e)] > **Ctrl+D** > > [Fig. 3.5(f)] > > **Remove Material** from the Dashboard > **Options** tab > Side 1 **Blind** > **Through All** > Side 2 **None** > **Through All** [Figs. 3.5(g-h)] > [Fig. 3.5(i)] > > **OK** > **File** > **Save As** > Type > **Zip File (*.zip)** > **OK** > **File** > **Close**

**Figure 3.5(d)** Select the Horizontal Surface

**Figure 3.5(e)** Sketch a Rectangle

**Figure 3.5(f)** Sketch Displayed in Standard Orientation

**Figure 3.5(g)** Through All

**Figure 3.5(h)** Preview of Cut

**Figure 3.5(i)** Completed Part

# Part Model Five (PRT0005.PRT) (Mirror)

Click: ▢ (prt0005) > **MMB** > ⬛ Extrude > press **RMB** > **Thicken Sketch** > pick datum **FRONT** > ⬛
> press **RMB** > **Construction Centerline** > pick two points vertically on the edge of datum RIGHT
[Figs. 3.6(a-b)] > move the mouse > **MMB** to end the current tool > **LMB** > **RMB** > **Line Chain** [Fig.
3.6(c)] > sketch the vertical and horizontal lines > **MMB** > **MMB** to display the dimensions [Fig. 3.6(d)]

**Figure 3.6(a)** Pick the First Point of the Vertical Centerline

**Figure 3.6(b)** Pick the Second Point of the Vertical Centerline

**Figure 3.6(c)** Line Chain    **Figure 3.6(d)** Sketch Two Lines

Click: **LMB** to deselect > press **RMB** > **Fillet** [Fig. 3.6(e)] > pick the lines near the corner > **MMB** to end the current tool > move the mouse pointer to the upper left corner of the Graphics Window > press **LMB** and hold while dragging a window until it incorporates the two lines and fillet [Fig. 3.6(f)] > release the **LMB** >  **Mirror the selected entities** > pick the Centerline [Fig. 3.6(g)] > **Ctrl+D** > ✓ [Fig. 3.6(h)] > 🔍 **Refit**

**Figure 3.6(e)** RMB > Fillet    **Figure 3.6(f)** Select Sketch Entities by Windowing

**Figure 3.6(g)** Mirrored Sketch

**Figure 3.6(h)** Depth Preview

Place the mouse pointer on the drag handle > press **RMB** > **Symmetric** [Fig. 3.6(i)] > move a drag handle [Fig. 3.6(j)] > > **LMB** to deselect > press and hold the **Ctrl** key and select the **RIGHT** and **FRONT** datum planes in the Model Tree > release the **Ctrl** key > **Axis** [Fig. 3.6(k)] > Hole > > > while holding down the **Ctrl** key, pick on the top face [Fig. 3.6(l)] > release the **Ctrl** key > **Enter** > **File** > **Save** > **OK** > **File** > **Save As** > Type > **Zip File (*.zip)** > **OK** > **File** > **Close**

Figure 3.6(i) Symmetric

Figure 3.6(j) Move a Drag Handle to Adjust the Depth

Figure 3.6(k) A_1 Axis Created

Figure 3.6(l) Hole Preview

## Part Model Six (PRT0006.PRT) (Revolve)

Click: 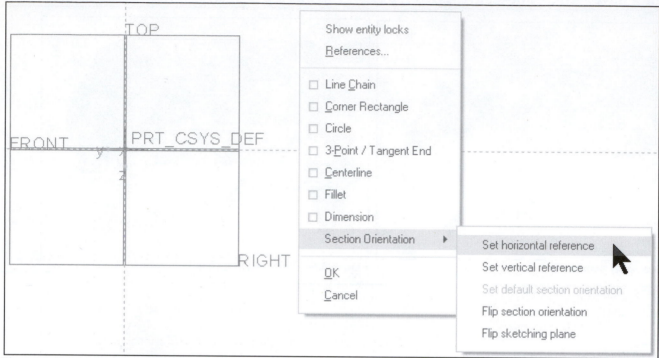 (prt0006) > **OK** > **View** tab > [icons] *on* > **Model** tab > pick datum **RIGHT** > **Revolve** > [icon] **Sketch View** > in the Graphics Window, press **RMB** > **Section Orientation** [Fig. 3.7(a)] > **Set horizontal reference** > pick on the Z-axis (edge of the TOP datum plane) [Figs. 3.7(b-c)]

**Figure 3.7(a)** Section Orientation

**Figure 3.7(b)** Select the Z-axis (Edge of Datum TOP)

**Figure 3.7(c)** Horizontal Reference Changed

Press: **RMB** > **Construction Centerline** > create a horizontal centerline (on the edge of the TOP datum) > move the mouse > **MMB** > press **RMB** > **Designate Axis of Revolution** [Fig. 3.7(d)] > press **RMB** > **Line Chain** > sketch the outline of the closed section > **MMB** to end the current line > **MMB** to end the current tool and to display the weak dimensions [Fig. 3.7(e)] > **Ctrl+D** > press **RMB** > **OK** [Fig. 3.7(f)]

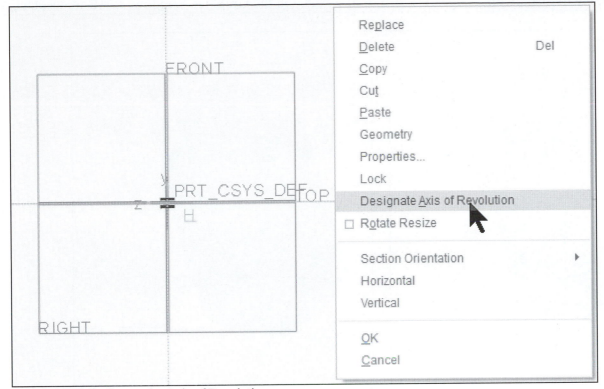

**Figure 3.7(d)** Designate the Axis of Revolution

**Figure 3.7(e)** Completed Revolved Sketch *(Note that there are only sketched entities on one side of the Axis of Revolution)*

Click:  [Fig. 3.7(g)] > in the Graphics Window, **LMB** to deselect > slowly pick on the front edge of the part > slightly move the mouse > pick on the front edge again until it highlights > press and hold the **Ctrl** key > pick the other visible edges > release the **Ctrl** key > press **RMB** > **Round Edges** [Fig. 3.7(h)] > Drag a handle to slightly adjust the radius value. The round surfaces will highlight if it is a valid size. (An invalid size will display only edges.) [Fig. 3.7(i)] > > in the Graphics Window, **LMB** to deselect

**Figure 3.7(f)** Revolve Preview

**Figure 3.7(g)** Completed Revolved Protrusion

**Figure 3.7(h)** Select the Edges

**Figure 3.7(i)** Move a Drag Handle to Adjust the Size

Click:  **View Manager** from the Graphics Toolbar > **Sections** tab > **New** [Fig. 3.7(j)] > **Planar** > *type* **A** > **Enter** > select datum **RIGHT** [Fig. 3.7(k)] > > `No Cross Section` > **RMB** > **Activate** [Fig. 3.7(l)] > **Close** > **File** > **Save** > **OK** > **File** > **Save As** > Type > **Zip File (*.zip)** > **OK** > **File** > **Close**

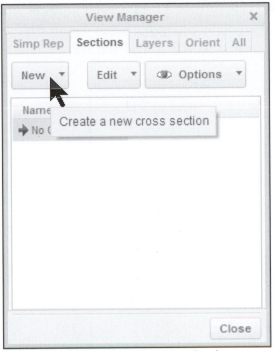

Figure 3.7(j) View Manager, New Section

Figure 3.7(k) Select the RIGHT Datum Plane

Figure 3.7(l) Section A

## Part Model Seven (PRT0007.PRT) (Revolve Ellipse)

Click: ⬜ (prt0007) > **OK** > pick datum **FRONT** > ⟨ Revolve > 🔁 **Sketch View** > in the Graphics Window, press **RMB** > **Axis of Revolution** > create a *vertical* Axis of Revolution on the edge of datum **RIGHT** > move the mouse > **MMB** > ⬭ ▾ > ⬭ **Center and Axis Ellipse** > pick a point to locate the center along the edge of the TOP datum plane > pick a point horizontally along the same datum edge > pick a third point to determine the shape of the ellipse [Fig. 3.8(a)] > **MMB** to end the current tool > **Ctrl+D** [Fig. 3.8(b)]

**Figure 3.8(a)** Sketch a Vertical Centerline and an Ellipse

**Figure 3.8(b)** Section 3D Preview

Click: [✓] > **Ctrl+D** > [360.00 ▼] > [180.00 ▼] > in the Graphics Window, press **RMB** > **Thicken Sketch** [Fig. 3.8(c)] > [✓] > **LMB** to deselect > **MMB** spin the model > [⊞] close the Model Tree *(lower left-hand corner below the Navigator)* > **Tools** tab > **Model** > [↔] to adjust the Browser shade [Fig. 3.8(d)] > [×] *(upper right corner of Browser)* or [⊘] *(lower left of screen)* to close the Browser > [⊞] open the Model Tree > [💾] > **OK** > **File** > **Save As** > Type [▼] > **Zip File (*.zip)** > **OK** > **File** > **Close**

**Figure 3.8(c)** Revolve Preview

**Figure 3.8(d)** Completed Elliptical Torus

## Part Model Eight (PRT0008.PRT) (Revolve Cut)

Press: **Ctrl+N** (prt0008) > **MMB** > pick datum **FRONT** > [Extrude] > [icon] > in the Graphics Window, press **RMB** > **Construction Centerline** > create a vertical centerline on the edge of datum RIGHT > move the mouse > **MMB** > **LMB** > press **RMB** > **Corner Rectangle** > sketch a symmetrical rectangle [icon] [Fig. 3.9(a)] > **MMB** > **Ctrl+D** > [✓] > [icon] > [icon] **Extrude on both sides** > pull one of the drag handles to increase the part's depth > **MMB** to complete the command > **LMB** to deselect

**Figure 3.9(a)** Sketch a Rectangle *(Notice the Symmetry Arrows)*

Click: [Revolve] > [icon] **Remove Material** from the Dashboard > [Placement] > [Define...] > [Use Previous] > [icon] > in the Graphics Window, press **RMB** > **References** > [icon] **Select references for dimensioning and constraining** > pick the top edge/surface of the part [Fig. 3.9(b)] > **Close**

**Figure 3.9(b)** Select the Top Edge/Surface as a Reference

Press: **RMB** > **Axis of Revolution** > create a *vertical* Axis of Revolution on the edge of datum RIGHT > move the mouse > **MMB** > **LMB** > press **RMB** > **Circle** > pick on the top edge > pick a radius position [Fig. 3.9(c)] > move the mouse > **MMB** [Fig. 3.9(d)] > **Ctrl+D** >  [Fig. 3.9(e)] > **MMB** [Fig. 3.9(f)]

**Figure 3.9(c)** Sketch a Circle

**Figure 3.9(d)** Completed Sketched Circle

**Figure 3.9(e)** Revolved Cut Preview

**Figure 3.9(f)** Completed Cut

Slowly pick on the edge of the part until it highlights > press and hold the **Ctrl** key > pick the other edges > release the **Ctrl** key > press **RMB** > **Round edges** [Fig. 3.9(g)] > Move a handle to adjust the radius value. Round surfaces will highlight for a valid size. [Fig. 3.9(h)] > **MMB** [Fig. 3.9(i)] > **LMB** to deselect

**Figure 3.9(g)** Select the Edges

**Figure 3.9(h)** Round Preview

**Figure 3.9(i)** Completed Round

**Figure 3.9(j)** Select Axis **A_1** _(your axis label may be different)_

**Figure 3.9(k)** Preview

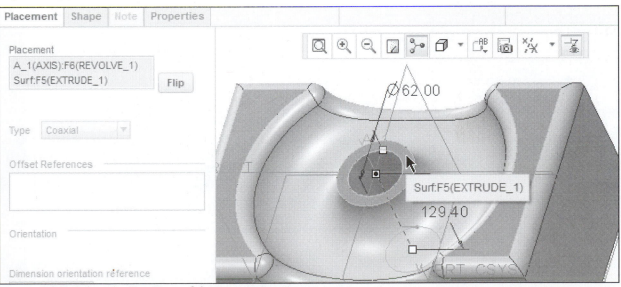

**Figure 3.9(l)** Select the Top Face of the Part

# Part Model Nine (PRT0009.PRT) (Blend)

Click: [icon] (prt0009) > **Enter** > **Shapes** group [Fig. 3.10(a)] > **Blend** > **Options** tab > [◉ Smooth] > **Sections** tab > **Define** [Fig. 3.10(b)] > pick datum **FRONT** > **Sketch** > [icon]

**Figure 3.10(a)** Blend Protrusion

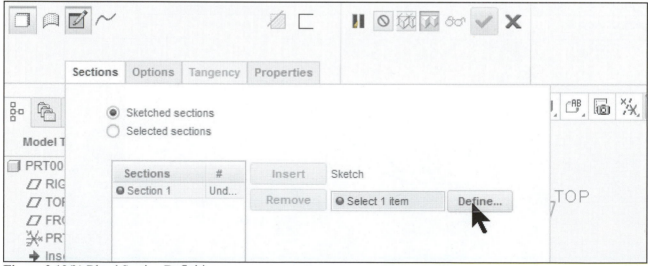

**Figure 3.10(b)** Blend Section Definition

Press **RMB** > **Construction Centerline** > create vertical and horizontal centerlines > move the mouse > **MMB** > **LMB** > press **RMB** > **Corner Rectangle** > start in the upper left-hand corner and pick two points (if you are careful, the rectangle will be symmetrical in both directions ) [Fig. 3.10(c)] > **MMB** > **LMB** > > > in the Graphics Window, press **RMB** > **Construction Centerline** > create vertical and horizontal centerlines again > move the mouse > **MMB** > **LMB** > press **RMB** > **Corner Rectangle** > again, start in the upper left-hand corner and pick two points [Fig. 3.10(d)] > **MMB** > > **Ctrl+D** >

**Figure 3.10(c)** First Section Sketched Rectangle *(Symmetric in Both Directions)*

**Figure 3.10(d)** Second Section Sketched Rectangle *(Symmetric in Both Directions)*

With the blend protrusion *still selected*, press: **RMB > Edit** [Fig. 3.10(e)] > place the mouse pointer on the drag handle [Fig. 3.10(f)] > press and hold down the **LMB** and move the drag handle [Fig. 3.10(g)] > release the **LMB** > move the mouse > **LMB** [Fig. 3.10(h)] > **Ctrl+D > Ctrl+S > Enter**

**Figure 3.10(e)** Edit

**Figure 3.10(f)** Depth Drag Handle

**Figure 3.10(g)** Move the Handle to Increase the Depth

**Figure 3.10(h)** Regenerated Object

If you have the license, click: **Flexible Modeling** tab >  **Move using Dragger** > pick on the front face [Fig. 3.10(i)] > press **LMB** on the *green* circle, drag to rotate > press **LMB** on the *red* circle, drag to rotate [Fig. 3.10(j)] > press **LMB** on the *blue* sphere and pull [Fig. 3.10(k)] > **Enter** > **Ctrl+S** > **OK** > **File** > **Save As** > Type  > **Zip File (*.zip)** > **OK** > **File** > **Close**

**Figure 3.10(i)** Set up Flexible Move

**Figure 3.10(j)** Rotate Face with Red Circle

**Figure 3.10(k)** Move Completed

145

## Part Model Ten (PRT0010.PRT) (Sweep)

For this part you will create the sketch before selecting a Shapes command. Press: **Ctrl+N** (prt0010) > OK > pick datum **FRONT** >  Sketch from the Model tab Ribbon > Sketch View > Spline Create a spline curve > starting at the part's origin, sketch a spline by picking 6 points [Fig. 3.11(a)] > **MMB** > **MMB** > **Ctrl+D** > > **File** > **Options** > **Model Display** > Default model orientation: **Trimetric** > **Isometric** > **OK** [Fig. 3.11(b)] > **No** > pick on the spline > move the mouse pointer slightly > pick on the spline again *(spline thickens)* > **Ctrl+S** > **Enter**

**Figure 3.11(a)** 6-point Spline

**Figure 3.11(b)** Select the Spline > Move the Mouse Pointer Slightly > Select the Spline Again

146

Click: [Sweep icon] Sweep [Fig. 3.11(c)] > pick on the arrow to toggle the start point to the origin of the spline [Fig. 3.11(d)] > [icon] > [icon] > [icon] **Corner Rectangle** [Fig. 3.11(e)] > pick two points of the rectangle *(symmetric in both directions)* > **MMB** [Fig. 3.11(f)]

**Figure 3.11(c)** Pick on the Arrow

**Figure 3.11(d)** New Starting Position

**Figure 3.11(e)** Sketch the Rectangle *(note symmetry)*

**Figure 3.11(f)** Dimensioned Sketch

Press: **Ctrl+D** [Fig. 3.11(g)] >  [checkmark] > [checkmark] [Fig. 3.11(h)] > *if the sweep does not display* > **Cancel** from the Regeneration Failure dialog box > **OK** from the Troubleshooter dialog box > [arrow]

**Figure 3.11(g)** Sweep Elements Displayed in 3D

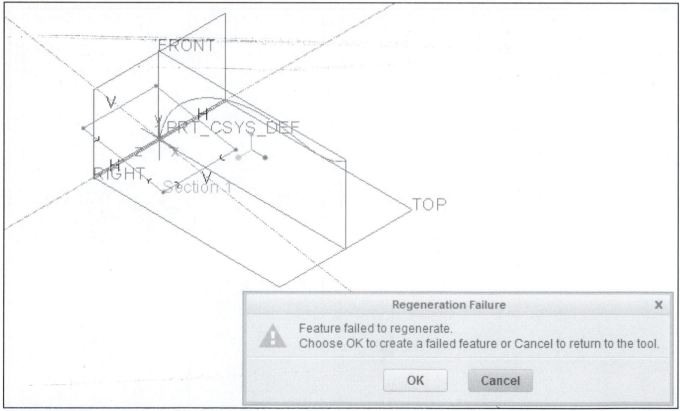

**Figure 3.11(h)** Regeneration Failure Dialog Box

Press: **LMB** on one corner of the rectangular sketch and drag inward until a preview of the sweep displays [Fig. 3.11(i)] > release the **LMB** > ☑ > in the Graphics Window, **LMB** to deselect > select **Sketch 1** from the Model Tree *(your Sketch number may be different)* > **RMB** > **Edit Definition** [Fig. 3.11(j)]

**Figure 3.11(i)** Feature Preview

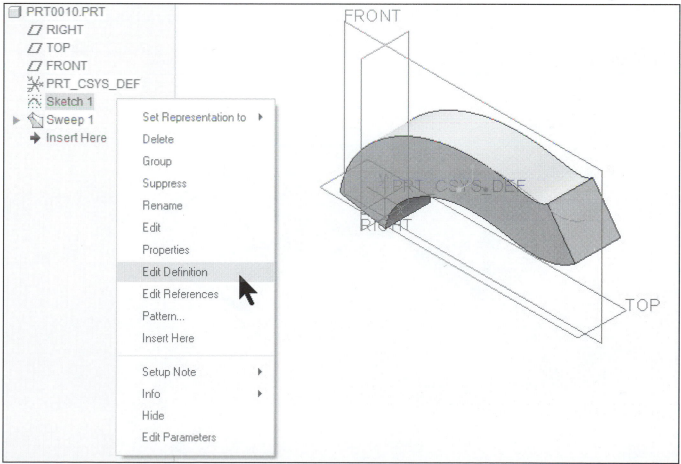

**Figure 3.11(j)** Edit Definition of the Sketch *(your Sketch number may be different)*

149

Press: **LMB** on each of the points of the spline and drag as desired [Fig. 3.11(k)] > press **RMB** > **OK** [Fig. 3.11(l)] > double-click on the sweep feature in the Graphics Window > press **LMB** on one of the rectangle's points and drag as desired [Fig. 3.11(m)] > release the **LMB** > move the pointer slightly >

**LMB** > **Ctrl+S** > **Enter** > **File** > **Save As** > Type  > **Zip File (\*.zip)** > **OK** > **File** > **Close** > **File** > **Exit** > **Yes** > on your computer; open Windows Explorer, locate and put all ten part zip files into one (new) folder > zip that new folder > upload or email the new zip file as directed by your instructor.

**Figure 3.11(k)** Drag each point to a new location    **Figure 3.11(l)** Edited Feature

**Figure 3.11(m)** Dynamically Edit the Section

**Figure 4.1** Clamp

## OBJECTIVES

- Create a feature using an **Extruded** protrusion
- Understand **Setup** and **Environment** settings
- Define and set a **Material** type
- Create and use **Datum** features
- Sketch protrusion and cut feature geometry using the **Sketcher**
- Understand the feature **Ribbon**
- **Copy** a feature

## REFERENCES AND RESOURCES

For **Resources** go to **www.cad-resources.com** > click on the Creo Parametric 2.0 Book cover

- Lesson 4 Lecture
- Lesson 4 3D models embedded in a PDF
- Book Projects PDF
- Project Lectures
- Creo Parametric 2.0 Quick Reference Card
- Creo Parametric 2.0 Configuration Options

## Extrusions

The design of a part using Creo Parametric 2.0 starts with the creation of base features (normally datum planes), and a solid protrusion. Other protrusions and cuts are then added in sequence as required by the design. You can use various types of Creo Parametric 2.0 features as building blocks in the progressive creation of solid parts (Fig. 4.1). Certain features, by necessity, precede other more dependent features in the design process. Those dependent features rely on the previously defined features for dimensional and geometric references. The progressive design of features creates these dependent feature relationships known as *parent-child relationships*. The actual sequential history of the design is displayed in the Model Tree. The parent-child relationship is one of the most powerful aspects of Creo Parametric 2.0 and parametric modeling in general. It is also very important as you modify a part. After a parent feature in a part is modified, all children are automatically modified to reflect the changes in the parent feature. It is therefore essential to reference feature dimensions so that Creo Parametric 2.0 can correctly propagate design modifications throughout the model.

A **protrusion** is a part feature that adds or removes material. A protrusion is *always the first solid feature created*. This is usually the first feature created after a base feature of datum planes. Figure 4.2 shows four common protrusions.

**Extrude**

**Revolve**

**Blend**

**Sweep**

**Figure 4.2** Basic Protrusions

## The Design Process

It is tempting to directly start creating models. Nevertheless, in order to build value into a design, you need to create a product that can keep up with the constant design changes associated with the design-through-manufacturing process. Flexibility must be integral to the design. Flexibility is the key to a friendly and robust product design while maintaining design intent, and you can accomplish it through planning. To plan a design, you need to understand the overall function, form, and fit of the product. This understanding includes the following points:

- Overall size of the part
- Basic part characteristics
- The way in which the part can be assembled
- Approximate number of assembly components
- The manufacturing processes required to produce the part

**Figure 4.3** Clamp and Datum Planes

## Clamp

The clamp in Figure 4.3 is composed of a protrusion and two cuts (one cut will be mirrored). A number of things need to be established before you actually start modeling. These include setting up the *environment*, selecting the *units*, and establishing the *material* for the part.

Before you begin any part, you must plan the design. The **design intent** will depend on a number of things that are out of your control and many that you can establish. Asking yourself a few questions will clear up the design intent you will follow: Is the part a component of an assembly? If so, what surfaces or features are used to connect one part to another? Will geometric tolerancing be used on the part and assembly? What units are being used in the design, SI or decimal inch? What is the part's material? What is the primary part feature? How should I model the part, and what features are best used for the primary protrusion (the first solid mass)? On what datum plane should I sketch to model the first protrusion? These and many other questions will be answered as you follow the systematic lesson part. However, you must answer many of the questions on your own when completing the *lesson project*, which does not come with systematic instructions.

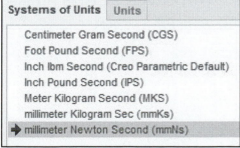

**Figure 4.4(a)** Units Manager Dialog Box

## Materials

| | |
|---|---|
| Material | Not assigned |
| Units | millimeter Newton Second (mmNs) |
| Accuracy | Relative 0.0012 |
| Mass Properties | |
| | Calculation source | Geometry |
| | Origin for calculation | PRT_CSYS_DEF |
| | Used density | 0.0 |

## Relations, Parameters and Instances

| | | |
|---|---|---|
| Relations | Not defined | |
| Parameters | 2 defined | |
| Instance | Not defined | Active: Generic - CLAMP |

## Features and Geometry

| | |
|---|---|
| Tolerance | ANSI |
| Names | 4 defined |

## Tools

| | | |
|---|---|---|
| Flexible | Not defined | |
| Shrinkage | Not defined | |
| Simplified Representation | 4 defined | Active: Master Rep |
| Pro/Program | | |
| Interchange | Not defined | |

## Model Interfaces

| | |
|---|---|
| Reference Control | Default settings |

## Detail Options

Detail Options

**Figure 4.4(b)** Model Properties Dialog Box

---

Click: Material **change** > **steel.mtl** > [▷▷▷] [Fig. 4.4(c)] > double-click on [➡STEEL] [Fig. 4.4(d)] > **Thermal** tab [Fig. 4.4(e)] > investigate other options and tabs > **Ok** > **OK** > **Close** > **Ctrl+S** > **Enter** *[you can end many commands by **Enter** or **OK** or **MMB** (Middle Mouse Button)]* > select [☐ CLAMP.PRT] in the Model Tree > **RMB** > **Info** > **Model**

| PART NAME : | CLAMP | ☐ |
|---|---|---|

MATERIAL FILENAME: STEEL

| Units: | Length: | Mass: | Force: | Time: | Temperature: |
|---|---|---|---|---|---|
| millimeter Newton Second (mmNs) | mm | tonne | N | sec | C |

> [×] close the Browser

**Figure 4.4(c)** Material File

**Figure 4.4(d)** Material Definition, Structural Tab

**Figure 4.4(e)** Material Definition, Thermal Tab

Since 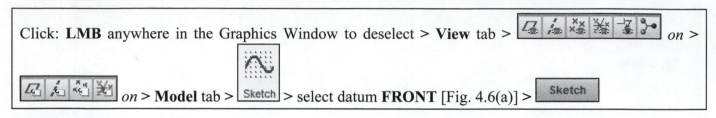 was selected, the default datum planes and the default coordinate system are displayed in the Graphics Window and in the Model Tree. *The **default datum planes** and the **default coordinate system** will be the first features on all parts and assemblies*. The datum planes are used to sketch on and to orient the part's features. Having datum planes as the first features of a part, instead of the first extrusion, gives the designer more flexibility during the design process. Picking on an item in the Model Tree will highlight/select that item on the model in the Graphics Window (Fig. 4.5).

Click: **LMB** anywhere in the Graphics Window to deselect > **View** tab > [icons] *on* >
[icons] *on* > **Model** tab > [Sketch] > select datum **FRONT** [Fig. 4.6(a)] > [Sketch]

**Figure 4.5** Default Datum Planes and Default Coordinate System

**Figure 4.6(a)** Sketch Dialog Box, Sketch Plane FRONT *(accept remaining default entries)*

Click:  **Sketch View** > in the Graphics Window, press **RMB** > **References** [Fig. 4.6(b)] *(the RIGHT and TOP datum planes are the positional/dimensional references)* > **Close** > **File** > **Options** > **Sketcher** > ☑ Show the grid > ☑ Snap to grid > **OK** > **No** [Fig. 4.6(c)] > **Shift+MMB** (pan) and **Ctrl+MMB** (zoom) to reposition and resize the sketch as needed.

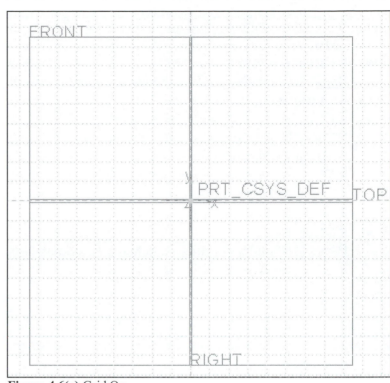

**Figure 4.6(b)** References Dialog Box

**Figure 4.6(c)** Grid On

The sketch is now displayed and oriented in 2D [Fig. 4.6(c)]. The coordinate system is at the middle of the sketch, where datum RIGHT and datum TOP intersect. The X coordinate arrow points to the right and the Y coordinate arrow points up. The Z arrow is pointing toward you (out from the screen). The square box you see is the limited display of datum FRONT. This is similar to sketching on a piece of graph paper. Creo Parametric 2.0 is not coordinate-based software, so you need not enter geometry with X, Y, and Z coordinates.

Before starting the sketch, spend some time investigating the Sketch Ribbon [Fig. 4.6(d)].

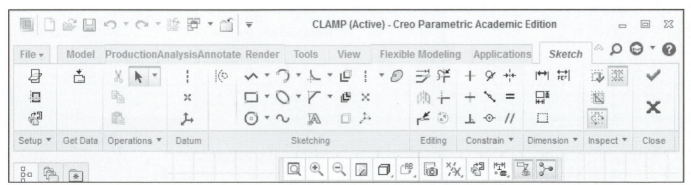

**Figure 4.6(d)** Sketch Ribbon *(shown here with "Hide Command Labels" checked, your Ribbon may appear differently)*

157

Because you checked 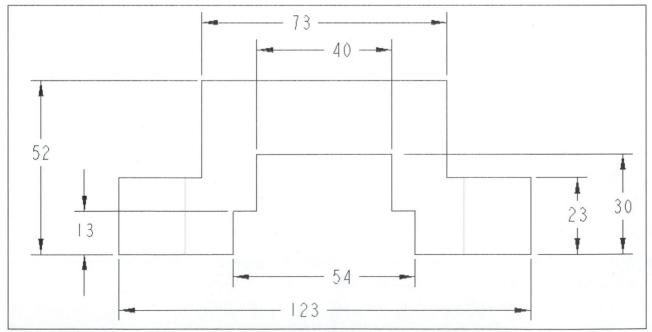 Snap to grid, you can now sketch by simply picking grid points representing the part's geometry (outline). Because this is a sketch in the true sense of the word, you need only create geometry that *approximates* the shape of the feature; the sketch does not have to be accurate as far as size or dimensions are concerned. No two sketches will be the same between those using these steps, unless you count each grid space (which is unnecessary). Even with the grid snap off, Creo Parametric 2.0 constrains the geometry according to rules, which include but are not limited to the following:

- **RULE:** Symmetry
  **DESCRIPTION:** Entities sketched symmetrically about a centerline are assigned equal values with respect to the centerline
- **RULE:** Horizontal and vertical lines
  **DESCRIPTION:** Lines that are approximately horizontal or vertical are considered exactly horizontal or vertical
- **RULE:** Parallel and perpendicular lines
  **DESCRIPTION:** Lines that are sketched approximately parallel or perpendicular are considered exactly parallel or perpendicular
- **RULE:** Tangency
  **DESCRIPTION:** Entities sketched approximately tangent to arcs or circles are assumed to be exactly tangent

The outline of the part's primary feature is sketched using a set of connected lines. The part's dimensions and general shape are provided in Figure 4.6(e). The cut on the front and sides will be created with separate sketched features. Sketch only one series of lines (8 lines in this sketch). ***Do not sketch lines on top of lines.***

It is important not to create any unintended constraints while sketching. Therefore, remember to exaggerate the sketch geometry and not to align geometric items that have no relationship. Creo Parametric 2.0 is very smart: if you draw two lines at the same horizontal level, Creo Parametric 2.0 assumes they are horizontally aligned. Two lines the same length will be constrained as so.

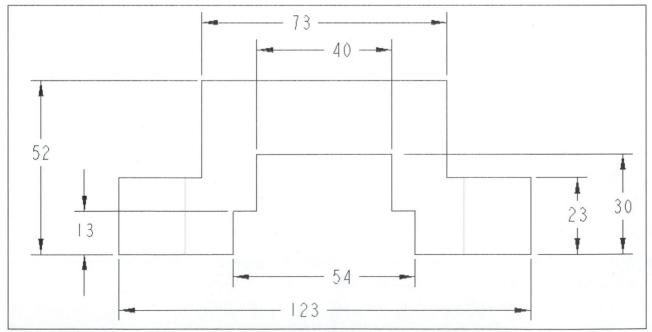

**Figure 4.6(e)** Front View of *Drawing* Showing Dimensions for the Clamp *(commands continue on the next page)*

158

With the cursor anywhere in the Graphics Window, but not on an object, press: **RMB** [Fig. 4.6(f)] > **Construction Centerline** > pick two vertical positions on the edge of datum RIGHT [Fig. 4.6(g)] > move the mouse > **MMB** > **LMB** >  > **Select All** Toggle all datum display filters *off* > **LMB** in the Graphics Window > press **RMB** > **Line Chain** > sketch eight lines *(watch the constraints as they appear-make each segment a different length and equal on one side of the centerline)* [Fig. 4.6(h)] > **MMB** to end the line sequence [Fig. 4.6(i)] > **MMB** > *If the sketch is not to your liking, click*  *and try again.*

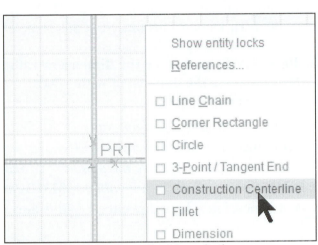

**Figure 4.6(f)** RMB Short Cut Menu Options

**Figure 4.6(g)** Create the Construction Centerline

**Figure 4.6(h)** Sketching the Closed Outline *(Do not make unwanted constraints. Line lengths should be different.)*

**Figure 4.6(i)** Default Dimensions Display *(Note the L1=L1 and L2=L2 constraints. Your sketch should appear similar.)*

A sketcher constraint symbol 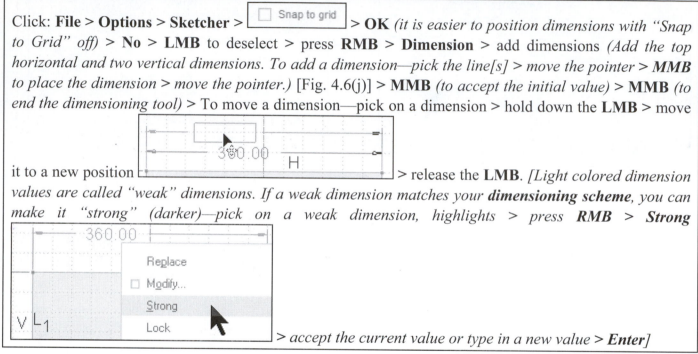 appears next to the entity that is controlled by that constraint. Sketcher constraints can be turned on or off (enabled or disabled) while sketching. Simply click the RMB as you sketch- before picking the position- and the constraint that is displaying will have a slash imposed over it. This will disable it for that entity. An H next to a line means horizontal; a T means tangent. Dimensions display, as they are needed according to the references selected and the constraints. Seldom are they the same as the required dimensioning scheme needed to manufacture the part. The dimensioning scheme is important, not the dimension value, which can be modified.

Place and create the dimensions as required. Do not be concerned with the perfect positioning of the dimensions, but in general, follow the spacing and positioning standards found in the **ASME Y14.5** standards. This saves you time when you create a drawing of the part. Dimensions placed at this stage of the design process are displayed on the drawing document by simply showing all the dimensions.

**To dimension between two lines, simply pick the lines (LMB) and place the dimension value (MMB). To dimension a single line, pick on the line (LMB), and place the dimension with MMB.**

Click: **File > Options > Sketcher >** □ Snap to grid **> OK** *(it is easier to position dimensions with "Snap to Grid" off)* **> No > LMB** *to deselect >* press **RMB > Dimension >** add dimensions *(Add the top horizontal and two vertical dimensions. To add a dimension—pick the line[s] > move the pointer >* **MMB** *to place the dimension > move the pointer.)* [Fig. 4.6(j)] **> MMB** *(to accept the initial value) >* **MMB** *(to end the dimensioning tool) >* To move a dimension—pick on a dimension > hold down the **LMB** > move

it to a new position > release the **LMB**. *[Light colored dimension values are called "weak" dimensions. If a weak dimension matches your **dimensioning scheme**, you can make it "strong" (darker)—pick on a weak dimension, highlights > press* **RMB > Strong**

> *accept the current value or type in a new value >* **Enter**]

**Figure 4.6(j)** Re-dimensioned Sketch *(you must have five similar dimensions- not the values)*

160

Next, control the sketch by adding symmetry constraints from the Sketch Ribbon; click: ⊹ **Symmetric** [Fig. 4.6(k)] > pick the centerline and then pick two vertices (endpoints) to be symmetric

[Fig. 4.6(l)] *(Note that because of symmetry, one of the weak dimensions is no longer needed and has been automatically deleted. Only four dimensions remain.)* > **File** > **Options** > **Sketcher** > ☐ Show the grid > **OK** > **No** [Fig. 4.6(m)]

Pick the centerline first, and then pick the two *endpoints* of the line

**Figure 4.6(k)** Constrain Palette          **Figure 4.6(l)** Adding Symmetry Constraint

Your original sketch *values* may be different from the example, but the final design values will be the same. ***DO NOT* CHANGE YOUR SKETCH DIMENSION VALUES TO THOSE IN FIGURE 4.6(m).**

**Figure 4.6(m)** Sketch is Symmetrical *(your values may be different! – DO NOT CHANGE YOUR VALUES AT THIS TIME!)*

**Again, do not change your *sketch* dimension values to these sketch values. Later, you will modify *your* sketch values to the required *design* dimensional values.**

161

Click: [Select] > Window-in the sketch (place the cursor at one corner of the window; with the **LMB** depressed, drag the cursor to the opposite corner of the window and release the **LMB**) to capture all four dimensions, which will *highlight*. > press **RMB** > **Modify** [Fig. 4.6(n)] > ☑ Regenerate *(checked by default)* > ☑ Lock Scale > resize the Modify Dimensions dialog box *(so that you can see all four dimensions)* [Fig. 4.6(o)] > in the Modify Dimensions dialog box, locate the dimension for the overall width *(here it is 600, but your dimension value may be different)*, double-click in that length dimension value field and *type* in the design value at the prompt (**123**) > **Enter**

**Figure 4.6(n)** RMB > Modify Dimensions

**Figure 4.6(o)** Modify the weak **600** Dimension to **123** *(your sketch weak dimension value may be different)*

Click:  **Regenerate the section and close the dialog** [Fig. 4.6(p)] > double-click on another dimension on the sketch (in the Graphics Window) and modify the value > **Enter** > continue until all of the values are changed to the design sizes [Fig. 4.6(q)] *(your sketch must have these dimensions and dimension values)* > in the Graphics Window, **LMB** to deselect

**Figure 4.6(p)** Modify each Dimension Individually

**Figure 4.6(q)** Modified Sketch showing the Design Values *(your sketch must have these dimensions and dimension values)*

Press: **Ctrl+D** [Fig. 4.6(r)] >  **Datum Display Filters** from the Graphics Toolbar > **Select All** Toggle all datum display filters *on* > **LMB** in the Graphics Window > **Sketch** tab > in the Graphics Window, press **RMB** > **OK** > 🖫 from the Quick Access Toolbar > **OK** *(or Enter or MMB)* {Sketch 1 remains *highlighted* in the Graphics Window and Model Tree, active and therefore *selected* [Fig. 4.6(s)].}

**Figure 4.6(r)** Regenerated Dimensions in the Standard Orientation View

**Figure 4.6(s)** Completed Sketched Curve *(Datum Curve) [your Sketch number may be different]*

With the sketch *still selected*, click: [ Extrude ] [Fig. 4.7(a)] > double-click on the depth value on the model > *type* **70** [Fig. 4.7(b)] > **Enter** > place the pointer over the square drag handle [ ] *(it will change color)* > press **RMB** > **Symmetric** [Fig. 4.7(c)] > **Enter** [Fig. 4.7(d)] > **Ctrl+S** > **Enter**

**Figure 4.7(a)** Depth of Extrusion Previewed

**Figure 4.7(b)** Modify the Depth Value

**Figure 4.7(c)** RMB > Symmetric

**Figure 4.7(d)** Completed Extrusion *(Sketch is hidden in the Model Tree)*

Click: **File > Options > Model Display > Trimetric > Isometric > OK > No > Ctrl+D > View** tab > **Display Style > Shading With Edges > Model** tab >  from the Graphics Toolbar > **Ctrl+S > Enter**

Storing an object on the disk does not overwrite an existing object file. To preserve earlier versions, Creo Parametric 2.0 saves the object to a new file with the same object name but with an updated version number. Every time you store an object using Save, you create a new version of the object in memory, and write the previous version to disk. Creo Parametric 2.0 numbers each version of an object storage file consecutively (for example, box.sec.1, box.sec.2, box.sec.3). If you save 25 times, you have 25 versions of the object, all at different stages of completion. You can use *File > Manage File > Delete Old Versions* [Fig. 4.7(e)] after the *Save* command to eliminate previous versions of the object that may have been stored.

When opening an existing object file, you can open any version that is saved. Although Creo Parametric 2.0 automatically retrieves the latest saved version of an object, you can retrieve any previous version by entering the full file name with extension and version number (for example, **partname.prt.5**). If you do not know the specific version number, you can enter a number relative to the latest version. For example, to retrieve a part from two versions ago, enter **partname.prt.3** *(or **partname.prt.-2**)*.

You use *File >Manage Session > Erase Current* to remove the object and its associated objects from memory. If you close a window before erasing it, the object is still in memory. In this case, you use *File >Manage Session > Erase Not Displayed* to remove the object and its associated objects from memory. This does not delete the object. It just removes it from active memory. *File >Manage File > Delete All Versions* removes the file from memory and from disk completely. You are prompted with a Delete All Confirm dialog box when choosing this command. Be careful not to delete needed files.

**Figure 4.7(e)** Delete Old Versions

166

Click: on Sketch 1 in the Model Tree > **RMB** > **Unhide** > Extrude [Fig. 4.8(a)] > **Placement** tab from the Dashboard > **Unlink** [Fig. 4.8(b)] > **OK** [Fig. 4.8(c)] > **Edit** [Fig. 4.8(d)]

**Figure 4.8(a)** Click on the Sketch in the Model Tree > Extrude

**Figure 4.8(b)** Unlink

**Figure 4.8(c)** Unlink Dialog Box

**Figure 4.8(d)** Internal Sketch Displayed

Click:  **Sketch View** > drag the end points and lines of the sketch away from the part's edges [Fig. 4.8(e)] > double-click on each value and modify to the design size [Fig. 4.8(f)] > **LMB** to deselect > **Ctrl+D**

**Figure 4.8(e)** Drag Sketch Lines and End Points

**Figure 4.8(f)** Modify Dimensions

Press: **RMB > OK** > press **RMB > Remove Material** > note the direction arrows [Fig. 4.8(g)] > **Options** tab from the Dashboard > Side 1 **Blind > Through All** > Side 2 **None > Through All** [Fig. 4.8(h)] > **MMB** to complete the feature > **Ctrl+S > Enter** [Fig. 4.8(i)] > **LMB** to deselect

**Figure 4.8(g)** Cut Preview *(note the direction arrows – your arrows may be at different locations)*

**Figure 4.8(h)** Options Depth

**Figure 4.8(i)** Completed Cut

The next feature will be a **20 X 20** centered cut (Fig. 4.9). Because the cut feature is identical on both sides of the part, you can mirror and copy the cut after it has been created.

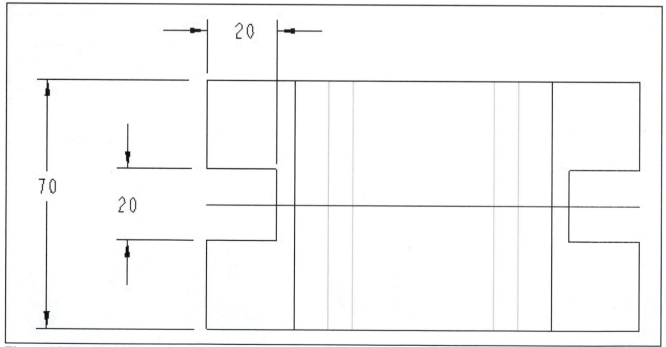

**Figure 4.9** Top View of *Drawing* Showing Dimensions for the Cut

Click:  Extrude > in the Graphics Window, press **RMB** > **Remove Material** > press **RMB** > **Define Internal Sketch** Sketch dialog box opens [Fig. 4.10(a)] > Sketch Plane--- Plane: select datum **TOP** from the model as the sketch plane > Sketch from the Sketch dialog box [Figs. 4.10(b-c)] >

**Figure 4.10(a)** Define Internal Sketch

**Figure 4.10(b)** Top Datum Selected as Sketch Plane

**Figure 4.10(c)** Sketch Dialog Box

171

In the Graphics Window, press: **RMB > References** > pick the left edge/surface of the part [Fig. 4.10(d)] to add it to the References dialog box [Fig. 4.10(e)] > **Close** > check to see if the grid snap is *off*, click: **File > Options > Sketcher > Cancel** *(if both "Show the Grid" and "Snap to Grid" are unchecked) [If both "Show the Grid" and "Snap to Grid" are checked, then uncheck them > OK > No]*

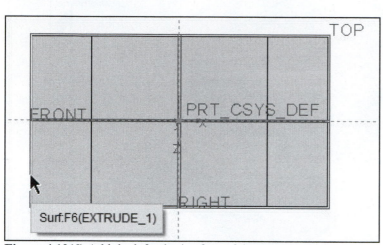

**Figure 4.10(d)** Add the left edge/surface of the part as a Reference

**Figure 4.10(e)** References Dialog Box

Click: **View** tab > **Display Style** > **Hidden Line** > **Sketch** tab > in the Graphics Window, press **RMB** > **Construction Centerline** [Fig. 4.10(f)] > create a *horizontal* centerline through the center of the part by picking two positions along the edge of the datum FRONT > move the mouse > **MMB** > **LMB**

**Figure 4.10(f)** Horizontal Centerline

Press: **RMB** > **Corner Rectangle** [Fig. 4.10(g)] > Place the mouse on the left edge, click **LMB** to begin creating a rectangle/square. Move the pointer and make sure the legs of the rectangle are equal: **L = L**. [Fig. 4.10(h)] > **LMB** to complete the current rectangle > **MMB** to end the rectangle tool [Fig. 4.10(i)] > double-click on the dimension [Fig. 4.10(j)] > *type* **20** > **Enter** [Fig. 4.10(k)] > move the mouse > **LMB** to deselect

**Figure 4.10(g)** RMB > Corner Rectangle

**Figure 4.10(h)** Sketch the Rectangle

**Figure 4.10(i)** Sketched Rectangle    **Figure 4.10(j)** Type 20    **Figure 4.10(k)** Completed Sketch

Press: **Ctrl+D** >  > from the Graphics Toolbar > **Shading With Edges** > **Change depth direction of extrude to other side of sketch** *[or click on the arrow (vertical in this case) in the Graphics Window]* > **Options** tab > Side 1 **Blind** > **Through All** [Fig. 4.10(l)] > in the Graphics Window, press **MMB** to spin the model [Fig. 4.10(m)] > move the mouse > release the **MMB** > from the Dashboard > > **Standard Orientation** > **File** > **Save** > **OK** [Fig. 4.10(n)]

**Figure 4.10(l)** Through All

Figure          4.10(m)          Cut          Preview

**Figure 4.10(n)** Completed Cut

With the cut still highlighted *[the extrude cut must be selected-highlighted for this tool to become active]*,

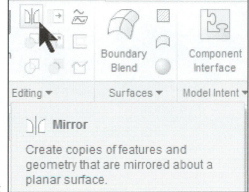

from the Ribbon click: **Mirror** > select datum **RIGHT** from the model in the Graphics Window *or in the Model Tree* [Fig. 4.10(o)] >  > in the Graphics Window, **LMB** (to deselect) > **Ctrl+S** > **OK** > **File** > **Save As** > Type > **Zip File (\*.zip)** > **OK**

**Figure 4.10(o)** With the Extruded Cut Highlighted (Selected), Select the RIGHT Datum Plane

175

Press: **MMB** spin the model >  **Show** in the Navigator > Expand All > in the Model Tree, select
next to Extrude 3 (2) to collapse > while pressing the **Ctrl** key, select **Extrude 3** and **Mirror 1** [Fig.
4.10(p)] > release the **Ctrl** key > select **CLAMP.PRT** in the Model Tree > **Ctrl+D** [Fig. 4.10(q)] > **File** >
**Save** > **OK** > **File** > **Manage File** > **Delete Old Versions** > **Enter** > **File** > **Close** > **File** > **Exit** > **Yes**

**Figure 4.10(p)** Spin the Model *(Extrude 3 and Mirror 1 have been selected in this view)*

**Figure 4.0(q)** CLAMP.PRT Selected in the Model Tree

A complete set of extra projects are available at ***www.cad-resources.com*** > click on the image of
your book cover > Lesson Projects.

# Lesson 5 Datums, Layers, and Sections

**Figure 5.1** Anchor Model with Datum Features hidden on a layer

## OBJECTIVES

- Create **Datums** to locate features
- Set datum planes for **geometric tolerancing**
- Learn how to change the **Color** and **Shading** of models
- Use **Layers** to organize features
- Use datum planes to establish **Model Sectioning**
- Add a simple **Relation** to control a feature
- Use **Info** command to extract **Relations** information

## REFERENCES AND RESOURCES

For **Resources** go to **www.cad-resources.com** > click on the Creo Parametric 2.0 Book cover

- Lesson 5 Lecture
- Lesson 5 3D models embedded in a PDF
- Book Projects PDF
- Project Lectures
- Creo Parametric 2.0 Quick Reference Card
- Creo Parametric 2.0 Configuration Options

## DATUMS, LAYERS, and SECTIONS

**Datums** and **layers** are two of the most useful mechanisms for creating and organizing your design (Fig. 5.1). Features such as *datum planes* and *datum axes* are essential for the creation of all parts, assemblies, and drawings using Creo Parametric 2.0.

Layers are an essential tool for grouping items and performing operations on them, such as selecting, hide/unhide, plotting, and suppressing. Any number of layers can be created. User-defined names are available, so layer names can be easily recognized.

Most companies have a layering scheme that serves as a *default standard* so that all projects follow the same naming conventions and objects/items are easily located by anyone with access. Layer information such as display status, is stored with each individual part, assembly, or drawing.

**Figure 5.2** Anchor

## Anchor

Though default datum planes have been sufficient in previous lessons, the Anchor (Fig. 5.2) incorporates the creation of user-defined datums and the assignment of datums to layers. The datum planes will be set as geometric tolerance features and put on a separate layer.

Launch **Creo Parametric 2.0** > **Select Working Directory** > navigate to your directory if necessary > **OK** > ☐ **New** > Name-- *type* **anchor** > ☑ Use default template > **OK** > **File** > **Options** > **Model Display** > Default model orientation: **Trimetric** > **OK** > **No** *(if needed)* > **File** > **Prepare** > **Model Properties** > Units **change** > **Inch lbm Second (Creo Parametric Default)** > **Close** > Material **change** > **steel.mtl** > ▶▶▶ > **OK** > **Close** > **View** tab > [icons] *on* > [icons] *on* > select the coordinate system on the model or in the Model Tree-- **PRT_CSYS_DEF** > press **RMB** [Fig. 5.3(a)] > **Rename** > *type* **csys_anchor** in the Model Tree > **Enter** *(automatically becomes uppercase)* [Fig. 5.3(b)]

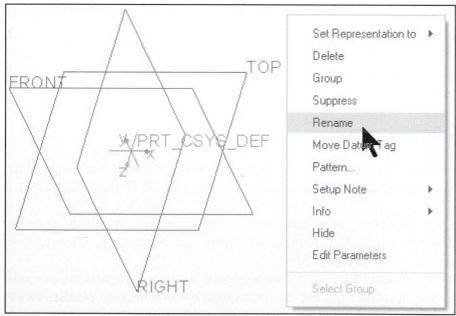

**Figure 5.3(a)** Rename the Default Coordinate System

**Figure 5.3(b)** Renamed CSYS

It is considered good practice to rename the default coordinate system to something similar to the components name. If an assembly has 25 components, it will have 25 default coordinate systems. Renaming the coordinate systems of each component will make them easier to visually identify.

The default datum planes and the default coordinate system are automatically placed on two layers each. For the datum planes; the *part default datum plane layer* and the *part all datum plane layer* [Fig 5.3(d)]. The coordinate system will also be layered in a similar fashion. The *part all datum coordinate system layer* for the coordinate system displays the new name created for the coordinate system.

Click: **Show** in the Navigator > **Layer Tree** Layer Tree displays in place of the Model Tree [Fig. 5.3(c)] > **Settings** > **Setup File** > **Save** [Fig. 5.3(d)] > **OK** > **Layers** *off* (from **View** tab Ribbon) > **Ctrl+S** > **OK** > in the Graphics Window, **LMB** to deselect

**Figure 5.3(c)** Show in the Navigator

**Figure 5.3(d)** Layers Displayed in Layer Tree

As in previous lessons, the first protrusion will be sketched on the FRONT datum plane. For the protrusion, use only the **5.50**, **R1.00**, **1.125** (vertical), and **25°** dimensions (Fig. 5.4).

**Figure 5.4** Anchor Drawing Front View

Click: **Model** tab > 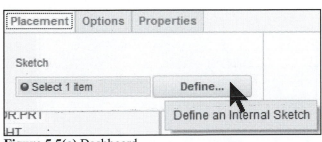 > **Placement** from the Dashboard > **Define** [Fig. 5.5(a)] > *in the Sketch dialog box:* Sketch Plane--- Plane: select datum **FRONT** from the model or Model Tree [Fig. 5.5(b)] > **Sketch** to close the Sketch dialog box > press **RMB** anywhere in the Graphics Window > **Line Chain**

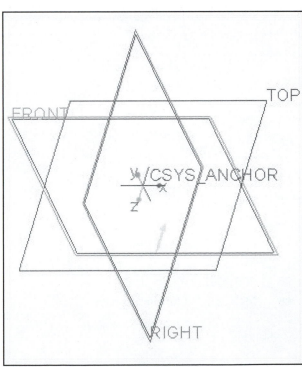

**Figure 5.5(a)** Dashboard

**Figure 5.5(b)** Select Datum FRONT

Starting at the coordinate system, sketch the four lines [Fig. 5.5(c)] > **MMB** > **MMB** > **LMB** > press **RMB** > **Fillet** > pick the left vertical line and the angled line to create a fillet [Fig. 5.5(d)] > **MMB** > **LMB** to deselect

**Figure 5.5(c)** Sketch Four Lines to Create a Closed Section > press RMB > Fillet

**Figure 5.5(d)** Create the Fillet (Dimensions are shown *weak* – your dimension values will be different)

Press: **RMB** > **Dimension** > pick on the horizontal line and then the angled line > move the pointer to the desired location for the dimension and click **MMB** to position the dimension value [Fig. 5.5(e)] > **Enter** *(or move the mouse > MMB)* to accept the initial value > **MMB** to end the current tool > reposition the dimensions > window-in dimensions *(or select all dimensions while pressing the **Ctrl** key)* > press **RMB** > **Modify** [Fig. 5.5(f)] > **Lock Scale** *on (if needed-- resize the Modify Dimensions dialog box)* > locate and modify the overall length dimension to **5.50** > **Enter** > from the Modify Dimensions dialog box

**Figure 5.5(e)** Create the Angle Dimension and Reposition the other Dimensions *(your dimension values will be different)*

**Figure 5.5(f)** After Capturing the Dimensions in a Window, Modify the Overall Length Dimension Value

Double-click on each dimension value and modify to the design size [Fig. 5.5(g)] > reposition dimensions if needed > **LMB** to deselect > press **RMB** > **OK** > *type* **2.5625** in the depth value field ⟨2.56250 ▾⟩ > **Enter** [Fig. 5.5(h)] > ✓ > **Ctrl+D** > **Ctrl+S** > **MMB** > in the Graphics Window, **LMB** to deselect

**Figure 5.5(g)** Modify the Dimensions to the Design Values

**Figure 5.5(h)** Modify the Depth to **2.5625** in the Dashboard

183

# Colors

**Colors** are used to define material and light properties. Avoid the colors that are used as Creo Parametric 2.0 defaults. Because feature and entity highlighting is currently defaulted to *green*, colors similar to *green* should be avoided. It is up to you to select colors that work with the type of project you are modeling. If the parts will be used in an assembly, each component should have a unique color scheme. In general, using light pastel colors would be better than using bright dark ones.

Click: **View** tab > Appearance Gallery ▼ [Fig. 5.6(a)] > **ptc-metallic-steel-light** [Fig. 5.6(b)] > 🖉 > ☐ ANCHOR.PRT ▱ RIGHT **ANCHOR.PRT** in the Model Tree > **OK** from the Select dialog box > Appearance Gallery ▼

**Appearance Gallery** > move the mouse over the Model Color ▼Model > **RMB** > **Edit** [Fig. 5.6(c)]

**Figure 5.6(a)** Appearance Gallery

**Figure 5.6(b)** ptc-metallic- steel-light

Click on the color button in the Model Appearance Editor dialog, Properties, Color 45.00 [Fig. 5.6(c)] *(color swatch)* to open the Color Editor dialog box [Fig. 5.6(d)] > move the **RGB/HSV Slider** bars to create a new color *(or type new values)* > you can also use the ▼ Color Wheel **Color Wheel** > or the ▼ Blending Palette **Blending Palette** > adjust the colors > **OK** from the Color Editor dialog box

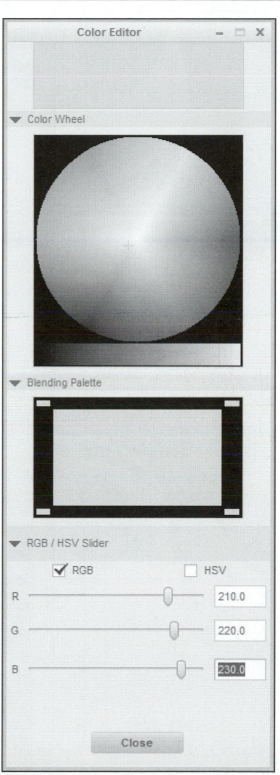

**Figure 5.6(c)** Model Appearance Editor    **Figure 5.6(d)** Color Editor

185

Click on the color button in the Model Appearance Editor dialog, Properties, Highlight Color ⬛
[Fig. 5.6(e)] *(color swatch)* > adjust the **RGB/HSV Slider** bars to create a new *highlight* color *(or type new values)* > ▼ Color Wheel **Color Wheel** > ▼ Blending Palette **Blending Palette > OK** from the Color Editor > adjust the Properties slide bars from the Model Appearance Editor [Fig. 5.6(f)] > **Close**

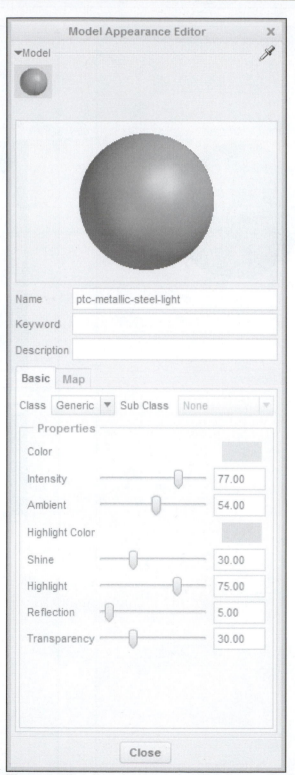

**Figure 5.6(e)** Highlight Color Editor

**Figure 5.6(f)** Adjust the Properties Slide Bars as Desired

Click: [Appearance Gallery ▼] > [Appearances Manager] **Appearances Manager** > [Fig. 5.6(g)] > *type* a New Name for the color **.dtm** file [New Name: CREO_Textbook, Type: Appearance (*.dmt)] > **OK** > **Close**

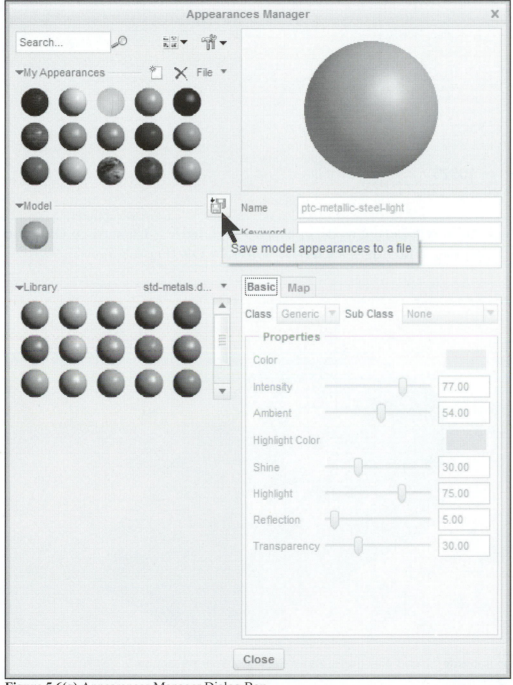

**Figure 5.6(g)** Appearances Manager Dialog Box

The next two features will remove material. For both of the cuts, use the RIGHT datum plane as the sketching plane, and TOP datum plane as the reference plane. Each cut requires just two lines, and two dimensions (Fig. 5.7). Use the first cut's sketching/placement plane and reference/orientation (**Use Previous**) for the second cut. ***Create the cuts separately***.

**Figure 5.7** Side View showing Dimensions for the Cuts

---

Click: **File** > **Options** > **Model Display** > Default model orientation: **Trimetric** > **Isometric** > **OK** > **No** > **View** tab > **Display Style** > **Shading With Edges** > **Ctrl+D** > **Ctrl+S** > **OK** [Fig. 5.8(a)]

---

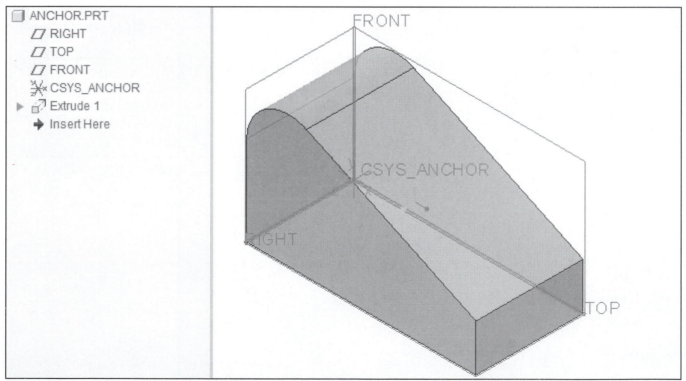

**Figure 5.8(a)** Standard Orientation Isometric, Shaded With Edges, Edited Color with Transparency on.

Click: **Model** tab > 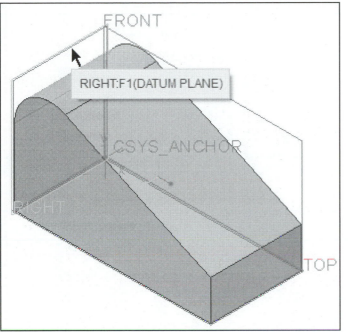 **Extrude** > in the Graphics Window, press **RMB** > **Remove Material** > press **RMB** > **Define Internal Sketch** > select datum **RIGHT** [Fig. 5.8(b)] for sketching [Fig. 5.8(c)] > **Sketch** to close the Sketch dialog box > **Sketch View** > in the Graphics Window, press **RMB** > **References** *(add a reference to align the cut with the top edge and the left edge of the part)* > pick on the *top edge/surface* of the part [Fig. 5.8(d)] > pick on the *left edge/surface* of the part [Fig. 5.8(e)] > **Solve** (notice the Reference status-- Fully Placed) [Fig. 5.8(f)] > **Close**

*(If you select the wrong edge/surface, pick the incorrect references in the References dialog box. > Delete > Select the two edges/surfaces again.)*

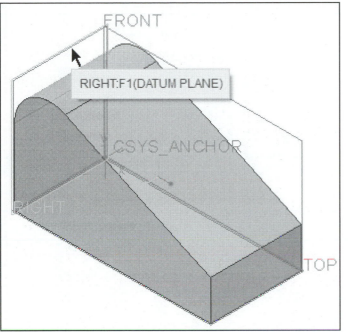

**Figure 5.8(b)** Select the Sketch Plane (RIGHT Datum Plane)　　**Figure 5.8(c)** Sketch Viewing Direction

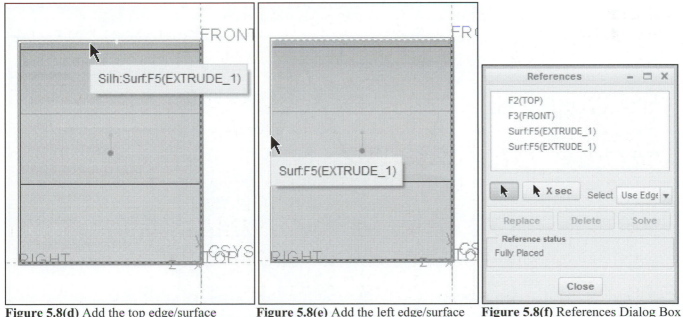

**Figure 5.8(d)** Add the top edge/surface　　**Figure 5.8(e)** Add the left edge/surface　　**Figure 5.8(f)** References Dialog Box

Press: **RMB** > **Line Chain** > sketch the two lines- start from the top edge with a *vertical* line and ending with the *horizontal* line locked to the left edge [Fig. 5.8(g)] > **MMB** > **MMB** [Fig. 5.8(h)] > **LMB** > press **RMB** > **Dimension** > *add* dimensions between the *horizontal* line and the bottom edge, and between the *vertical* line and the right edge [Fig. 5.8(i)], *pick on the lines and edges not the endpoints* > **MMB** (place the last dimension) > move the mouse > **MMB** (accept the default value) > **MMB** (end the current tool)

*Note: "weak" (lighter colored) dimensions disappear as new dimensions are added. Created dimensions are "strong" dimensions (darker colored) and represent the design intent of the feature. These dimensions can be displayed on a drawing as driving-design dimensions to be used in manufacturing.*

**You must have the dimensioning *scheme* shown in Figure 5.8(i).** Your dimension values will be different. Do not modify your dimensions to the ones shown here.

**Figure 5.8(g)** Sketch a Vertical and a Horizontal Line

**Figure 5.8(h)** Default Dimensions *(weak)* Displayed

**Figure 5.8(i)** New Dimensions *(strong)* (*You* must have the dimensioning *scheme* shown here- *your values will be different*)

190

The dimensional *design scheme* is correct for the cut, but the values are not the correct *design values*. Double-click on the *horizontal* dimension and change to the design value. [Fig. 5.8(j)] > **Enter** > repeat for the *vertical* dimension [Fig. 5.8(k)] > **LMB** in Graphics Window > **Ctrl+D** > press **RMB** > **OK** >

place the pointer on the depth direction arrow  > **LMB** to select *(flips the depth direction arrow to the other side)* [Fig. 5.8(l)]

**Figure 5.8(j)** Modify Dimension to **1.875**

**Figure 5.8(k)** Modify Dimension to **1.125**

**Figure 5.8(l)** Extrusion Preview *(flip arrows if necessary)*

191

Place the pointer over the drag handle > press **RMB** [Fig. 5.8(m)] > **Through All** [Fig. 5.8(n)] > **Options** tab > **MMB** to complete the feature >  **Refit** > in the Model Tree, click: ▶ (to expand) > ▶ > select [Extrude 2 / Section 1] [Fig. 5.8(o)] > in the Graphics Window, **LMB** to deselect > 💾 **Save** > **OK**

**Figure 5.8(m)** RMB on Drag Handle          **Figure 5.8(n)** Through All

**Figure 5.8(o)** Completed Cut

Click: **File** > **Options** > **Sketcher** > Number of decimal places for dimension: 2 > **4** > **OK** > **Yes** > **OK** > Extrude > Remove Material from the Dashboard > in the Graphics Window, press **RMB** > **Define Internal Sketch** > **Use Previous** > Sketch View > in the Graphics Window, press **RMB** > **References** > add the left *vertical* edge/surface as a reference [Fig. 5.9(a)] > **Close** > Line Chain > sketch the vertical and then the horizontal lines > **MMB** > **MMB** [Fig. 5.9(b)] > **LMB** > reposition the default weak dimension scheme by moving the pointer on top of each dimension value, press and hold down the **LMB**, move the pointer to a new position, release the **LMB** to place the dimension [Fig. 5.9(c)]

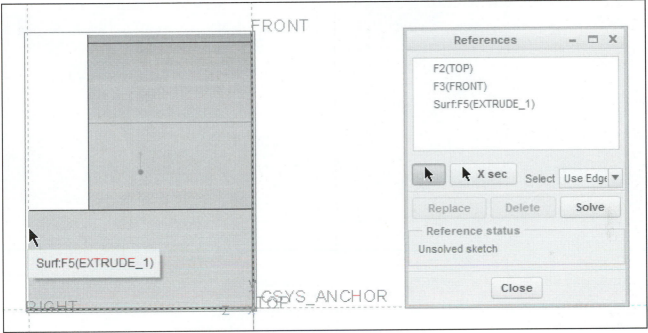

Figure 5.9(a) Add the Left Vertical Edge/Surface as a Reference

Figure 5.9(b) Sketch the Two Lines

Figure 5.9(c) Weak Dimensions

193

Double-click on the vertical dimension and *type* **.5625** [Fig. 5.9(d)] > **Enter** > double-click on the horizontal dimension and *type* **1.0625** [Fig. 5.9(e)] > **Enter** > **Ctrl+D** > ☑ from the Ribbon > ⬛ > ⬛ **Extrude to intersect with all surface** > flip the depth direction arrow if necessary [Fig. 5.9(f)] > ☑ from the Dashboard [Fig. 5.9(g)] > **Ctrl+S** > **OK** > **LMB** to deselect

**Figure 5.9(d)** Modify Vertical Dimension to **.5625**

**Figure 5.9(e)** Modify Horizontal Dimension to **1.0625**

**Figure 5.9(f)** Dynamic Preview of the Cut

**Figure 5.9(g)** Completed Cut

For the next feature, you will need to create a new datum plane on which to sketch. New part datum planes are by default numbered sequentially, DTM1, DTM2, and so on. Also, a datum axis (A_1) will be created through the curved top of the part and used later for the axial location of a small hole. Holes and circular features automatically have axes when they are created. Features created with arcs, fillets and so on need to have an axis added if needed as a design reference.

Click: **Datum Display Filters** from the Graphics Toolbar > ☐ Csys Display *off* > **LMB** in the Graphics Window > Plane from the Ribbon Datum Group > References: pick datum **FRONT** as the plane to offset from [Fig. 5.10(a)] > in the DATUM PLANE dialog box, Offset: Translation field, *type the distance* **1.875/2** *(1.875/2 = .9375, always have the CAD system do the math!)* > **Enter** *(make sure that the value is .9375)* > **OK** > **LMB** to deselect > Axis > References: pick the curved surface [Fig. 5.10(b)] > **OK** > **LMB** to deselect *(you must deselect before the next feature is created)* > **Ctrl+S** > **OK**

**Figure 5.10(a)** Offset Datum

**Figure 5.10(b)** Datum Axis A_1 through the Center of the Curved Surface

195

Click: [⬜ Plane] > References: pick the angled surface > in the References field, click: **Offset** to activate > **Offset** to see options > **Through** [Fig. 5.10(c)] > **OK** > [⬜] > [☑ ⬜ Csys Display] *on* [Fig. 5.10(d)] > **LMB** in the Graphics Window > [💾] > **OK** > **LMB** to deselect

**Figure 5.10(c)** Select the Angled Surface, then select Through

**Figure 5.10(d)** Through Datum Plane

196

Create a new layer and add the new datum planes and axis to it, click: **View** tab > ▤ Layers *(Layer Tree displays in place of the Model Tree in the Navigator)* > ▶ > ▶ [Fig. 5.11(a)] > Layer Tree ▤ > **New Layer** [Fig. 5.11(b)] > In the Layer Properties dialog box [Fig. 5.11(c)], Name: *type* **DATUM_FEATURES** *(do not press Enter)* as the name for new layer [Fig. 5.11(d)]

**Figure 5.11(a)** Layer Tree in the Navigator

**Figure 5.11(b)** New Layer

**Figure 5.11(c)** Layer Properties Dialog Box

**Figure 5.11(d)** Layer Name - DATUM_FEATURES

Select the two *new* datum planes from the model > with the cursor over the axis on the model, click **RMB** to toggle to the correct selection *(axis will highlight)* > select the axis with **LMB** [Fig. 5.11(e)] > **OK**

[Fig. 5.11(f)] > ▶ expand layer to see items [Fig. 5.11(g)] > in the Navigator > **Model Tree** > **Ctrl+R** *(repaints/redraws the screen)* > **Ctrl+S** > **OK**

▼ ⬭ DATUM_FEATURES
⬜ F8(DTM1)
⁄ F9(A_1)
⬜ F10(DTM2)

**Figure 5.11(e)** Highlight Axis using RMB, then LMB to Select

**Figure 5.11(f)** Adding Items to a Layer

**Figure 5.11(g)** DATUM_FEATURES Layer

# Geometric Tolerances

Before continuing with the modeling of the part, the datums used and created thus far will be "set" for geometric tolerancing. Geometric tolerances (GTOLs) provide a comprehensive method of specifying where on a part the critical surfaces are, how they relate to one another, and how the part must be inspected to determine if it is acceptable. They provide a method for controlling the location, form, profile, orientation, and run out of features. When you store a Creo Parametric 2.0 GTOL in a solid model, it contains parametric references to the geometry or feature it controls—its reference entity—and parametric references to referenced datums and axes. In Assembly mode, you can create a GTOL in a subassembly or a part. A GTOL that you create in Part or Assembly mode automatically belongs to the part or assembly that occupies the window; however, it can refer only to set datums belonging to that model itself, or to components within it. It cannot refer to datums outside of its model in some encompassing assembly, unlike assembly created features. You can add GTOLs in Part or Drawing mode, but they are reflected in all other modes. Creo Parametric 2.0 treats them as annotations, and they are always associated with the model. Unlike dimensional tolerances, though, GTOLs do not affect part geometry. Before you can reference a datum plane or axis in a GTOL, you must set it as a reference. Creo Parametric 2.0 encloses its name using the set datum symbol. After you have set a datum, you can use it in the usual way to create features and assemble parts. You enter the set datum command by picking on the datum plane or axis in the Model Tree or on the model itself > RMB > Properties. Creo Parametric 2.0 encloses the datum name in a feature control frame. If needed, type a new name in the Name field of the Datum dialog box. Most datums will follow the alphabet, A, B, C, D, and so on. You can hide (not display) a set datum by placing it on a layer and then hiding the layer *(or selecting on the item in the Model Tree > RMB > Hide)*.

Select datum **TOP** in the Model Tree > **RMB** > **Properties** [Figs. 5.12(a-b)]

**Figure 5.12(a)** Datum Properties

199

Double-click in the Name field > *type* **A** > **Set** [Fig. 5.12(c)] > **OK** in the Datum dialog box > repeat the process and set datum FRONT as **B** and datum RIGHT as **C**. You can also set datums directly on the model. Pick **DTM1** on the model > press **RMB** > **Properties** > Name- *type* **D** > **Set** > **OK** > pick **DTM2** > press **RMB** > **Properties** > Name- *type* **E** > **Set** > **OK** [Fig. 5.12(d)] > **Ctrl+S** > **OK** > **LMB**

**Figure 5.12(b)** Datum Dialog Box

**Figure 5.12(c)** Set

**Figure 5.12(d)** Datums Set on the Model

Create the slot on the top of the part by sketching on datum D and projecting the cut toward both sides. Use the Model Tree to select the appropriate datum planes.

Click: **Model** tab > [Extrude] > in the Graphics Window, press **RMB** > **Define Internal Sketch** > Plane: select datum **D** [Fig. 5.13(a)] > **Sketch** > add a references to align the cut with the angled edge of the part, click: [References icon] **References** from the Ribbon Setup Group > pick on datum **E** > pick the small *vertical* surface on the right side of the part as the fourth reference [Fig. 5.13(b)] > **Close**

**Figure 5.13(a)** Sketch Plane and Sketch Orientation Reference

**Figure 5.13(b)** New References (Datum E and Small Right Vertical Surface)

201

Click:  *off* > > Hidden Line > **Sketch View** > **Point** from the Sketch Ribbon Sketching Group > pick at the corner of the angled datum **E** and the right vertical reference [Fig. 5.13(c)] > move the mouse > **MMB** to end the current command > **LMB** to deselect > **Line Chain** from the Sketch

Ribbon Sketching Group > sketch the first line from and perpendicular to the angled edge [Fig. 5.13(d)]

**Figure 5.13(c)** Construction Point

**Figure 5.13(d)** Create the first Line Perpendicular to Datum E

Sketch the next line horizontal (to the left vertical reference) [Fig. 5.13(e)] > **MMB** > **MMB** to end the line command [Fig. 5.13(f)] > **LMB** to deselect > move the dimension off of the object [Fig. 5.13(g)]

**Figure 5.13(e)** Create the Second Line Horizontal

**Figure 5.13(f)** Completed Lines

**Figure 5.13(g)** Move the Dimension

Double-click on the moved dimension > *type the design value of* **1.50** [Fig. 5.13(h)] > **Enter** > move the mouse off of the dimension > **LMB** to deselect > press **RMB** > **Dimension** > add a dimension by selecting angled line and then the construction point *(do not pick endpoint to point)* > **MMB** to place the dimension [Fig. 5.13(i)] > *type* **2.625** > **Enter** > **MMB** to end the dimensioning tool > **Ctrl+D** > ✔

**Figure 5.13(h)** Modify the Vertical Dimension to **1.500**

**Figure 5.13(i)** Create Dimension from First Line to Construction Point (not end point to point!)

The **2.625** value is different than that shown in Figure 5.4. When this part is used in a later lesson, the drawing detail will show the dimension modeled here. You will be instructed on how to modify the value at the drawing level instead of on the model. Since the part, assembly and drawing are associative, the component will regenerate with the new size.

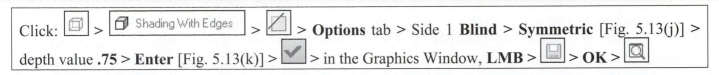

Click: [icon] > [Shading With Edges] > [icon] > **Options** tab > Side 1 **Blind** > **Symmetric** [Fig. 5.13(j)] > depth value **.75** > **Enter** [Fig. 5.13(k)] > [✓] > in the Graphics Window, **LMB** > [icon] > **OK** > [icon]

**Figure 5.13(j)** Symmetric Cut Preview

**Figure 5.13(k)** Symmetric Depth Value

The hole drilled in the angled surface appears to be aligned with datum **D**. Upon inspection (Fig. 5.7), the hole is at a different distance and is not in line with the slot and datum plane.

Click: [ Hole ] > pick on the angled face and a hole will display with drag handles [Fig. 5.14(a)] > drag and drop one *diamond* handle to datum **B** [Fig. 5.14(b)] and the other to the edge between the angled surface and the right vertical face [Fig. 5.14(c)] > **RMB** anywhere inside the Graphics Toolbar > **Location** > **Show on the Bottom**

**Figure 5.14(a)** Hole Placement (two diamond placement drag handles and three square drag handles)

**Figure 5.14(b)** Drag and Drop Handle to Datum B

**Figure 5.14(c)** Drag and Drop other Handle to the Edge

Double-click on the dimension from the hole's center to datum **B** > *type* **.875** [Fig. 5.14(d)] > **Enter** > double-click on the dimension from the hole's center to the edge > *type* **2.0625** [Fig. 5.14(e)] > **Enter** >  **Use standard hole profile as drill hole profile** in the Dashboard > ⌀ 1.0000 > 1.1250 > **Shape** tab [Fig. 5.14(f)] > ✓ > **Ctrl+D** > in the Graphics Window, **LMB** to deselect

**Figure 5.14(d)** Modify the Distance to Datum B

**Figure 5.14(e)** Modify the Distance to the Part Edge

**Figure 5.14(f)** Shape Tab

207

Select axis **A_1** from the Model Tree > [ Hole ] > hold down the **Ctrl** key and pick datum **D** from the model or the Model Tree > release the **Ctrl** key > **Placement** tab [Fig. 5.15(a)] > **Shape** tab > diameter **.250** [ Ø 0.25 ] > **Enter** > **Blind** > **Through All** > Side 2 **None** > **Through All** [Fig. 5.15(b)] > **MMB** > **Ctrl+D** > **Ctrl+S** > **Enter** > **Ctrl+R** > in the Graphics Window, **LMB** to deselect

**Figure 5.15(a)** Placement Tab Placement References

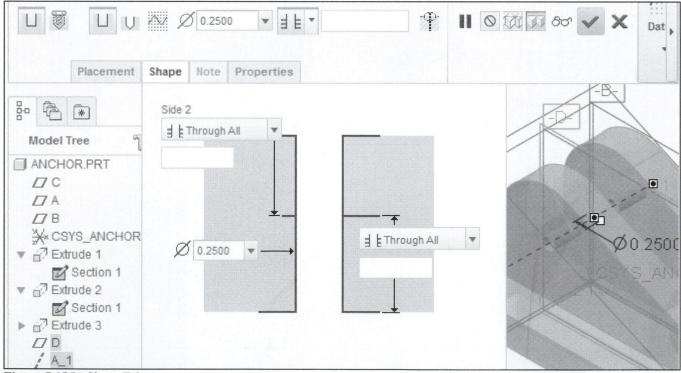

**Figure 5.15(b)** Shape Tab

## Suppressing and Resuming Features using Layers

Next, you will create a new layer and add the two holes to it. The holes will then be selected in the Layer Tree and suppressed. Suppressed features are temporarily removed from the model along with their children (if any). In the example, you will notice that the holes and one axis will be suppressed. Since axis A_1 is the parent of the small hole, it will not be suppressed. Suppressing features is like removing them from regeneration temporarily. However, you can resume suppressed features as needed.

You can suppress features on a part to simplify the part model and decrease regeneration time. For example, while you work on one end of a shaft, it may be desirable to suppress features on the other end of the shaft. Similarly, while working on a complex assembly, you can suppress some of the features and components for which the detail is not essential to the current assembly process. Suppress features to do the following:

- Concentrate on the current working area by suppressing other areas
- Speed up a modification process because there is less to update
- Speed up the display process because there is less to display
- Temporarily remove features to try different design iterations

Unlike other features, the base feature cannot be suppressed. If you are not satisfied with the base feature, you can redefine the section of the feature, or start another part.

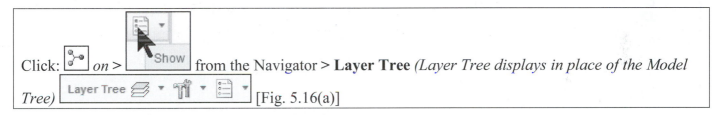

Click: [icon] *on* > Show from the Navigator > **Layer Tree** *(Layer Tree displays in place of the Model Tree)* [Layer Tree icons] [Fig. 5.16(a)]

**Figure 5.16(a)** Layer Tree

Click:  Layers > **RMB** > **New Layer** [Fig. 5.16(b)] > in the Layer Properties dialog box, Name: *type* **HOLES** as the name for the new layer > **Enter** > select the new **HOLES** layer in the Layer Tree > **RMB** > **Layer Properties** [Fig. 5.16(c)] *(you can avoid doing this by **not** pressing Enter after you type in a new layer name)* > Selection Filter (the lower right-hand side below the Graphics Window), click: **All** > **Feature** [Fig. 5.16(d)] > select the two holes from the model *(no need to hold down the Ctrl key)* [Fig. 5.16(e)] > **OK** > **LMB** in the Graphics Window > expand the HOLES layer > press and hold down the **Ctrl** key and pick on the two holes in the HOLES layer [Fig. 5.16(f)] > release the **Ctrl** key

**Figure 5.16(b)** New Layer

**Figure 5.16(c)** RMB > Layer Properties

**Figure 5.16(d)** Filter

**Figure 5.16(e)** Holes Layer Items

**Figure 5.16(f)** Holes Selected

Click:  **Command Search** > *type* **suppr**ess > in the list, place the mouse on the icon (but do not click the **LMB**) [Suppress(1)] [Fig. 5.16(g)] > [🔍] to close > **Model** tab > **Operations** Group > [▶] next to Suppress > **Suppress** [Fig. 5.16(h)] > **OK** > **LMB** to deselect *(look at the Layer Tree and the Model to see that the holes are gone)* > [📄] > **Model Tree** > [📄] > **Layer Tree** > **Operations** Group > **Resume** > **Resume All** > [📄] > **Model Tree** > in the Graphics Window, **LMB** to deselect [Fig. 5.16(i)] > [💾] > **OK**

**Figure 5.16(g)** Command Search > *type* suppr

**Figure 5.16(h)** Suppress

**Figure 5.16(i)** Resume All

# Cross Sections

There are six types of cross sections: **Planar, X Direction**, **Y Direction**, **Z Direction**, **Offset**, and **Zone**. You will be creating a planar cross section. Creo Parametric 2.0 can create standard planar cross sections of models (parts or assemblies), offset cross sections of models (parts or assemblies), planar cross sections of datum surfaces or quilts (Part mode only) and planar cross sections that automatically intersect all quilts and all geometry in the current model.

---

Click: **View** tab > ![icon] **View Manager** View Manager dialog box displays [Fig. 5.17(a)] > **Sections** tab [Fig. 5.17(b)] > **New** [Fig. 5.17(c)] > **Planar** [Fig. 5.17(d)] > *type* **A** [Fig. 5.17(e)] > **Enter** > ![icon] [Fig. 5.17(f)] > in the Graphics Window, press **RMB** > **Section Reference** [Fig. 5.17(g)]

---

**Figure 5.17(a)** View Manager

**Figure 5.17(b)** Sections Tab

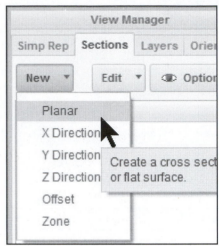

**Figure 5.17(c)** New > Planar

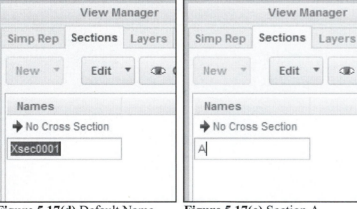

**Figure 5.17(d)** Default Name    **Figure 5.17(e)** Section A

**Figure 5.17(f)** Activate Dragger

**Figure 5.17(g)** Section Reference

212

Select datum **D** [Fig. 5.17(h)] >  > select a color [Fig. 5.17(i)] > > > > > > > **RMB** on > **Edit Hatching** > **iso_alloy_steel** > **Apply** > **steel** > **Apply** > Angle field, *type* **150** > **Enter** > **Use a solid fill** > **Use hatch from the part** > **Apply** > > **RMB** on **No Cross Section** > **Activate** [Fig. 5.17(j)] > **Close** > in Model Tree, **RMB** on > uncheck **Show Section** > in the Graphics Window, **LMB** to deselect > **File** > **Save** > **OK**

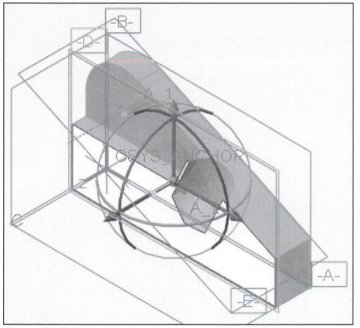

**Figure 5.17(h)** Section Preview

**Figure 5.17(i)** Section Color

**Figure 5.17(j)** Activate No Cross Section

## Relations

The section passes through datum D. The slot is the *child* of datum D. If datum D moves, so will the slot and the small hole (and X-Section A). In order to ensure that datum D stays centered on the upper portion of the part, you will need to create a relation to control the location of datum D. **Relations** will be covered in-depth in a later lesson. The first cut used a dimension from datum B for location (**1.875**). The relation should state that the distance from datum B to datum D will be one-half the value of the distance from datum B to the first cut. If the thickness of the upper portion of the part (**1.875**) changes, datum D will remain centered as will the slot and the X-Section. To start, you must first determine the feature dimension symbols (**d#**) required for the relation.

**(READ! - Almost everyone will have different d# numbers than the ones displayed here)**

---

Select the first cut (extrude) from the Model Tree > **RMB** > **Edit** [Figs. 5.18(a-b)] > **Tools** tab > [Switch Symbols] > note symbol **d6** *(Note: your "d" values may be different)* [Fig. 5.18(c)] > in the Graphics Window, **LMB** to deselect > select datum **D** in the Model Tree > **RMB** > **Edit** > note symbol **d11** [Fig. 5.18(d)] *(Note: your "d" values may be different. If you have the wrong symbols, your relation will not work. **Use your d values not the ones in the text!**)* > in the Graphics Window, **LMB** to deselect

---

**Figure 5.18(a)** Edit  **Figure 5.18(b)** Dimension  **Figure 5.18(c)** d6 Symbol

**Figure 5.18(d)** Using Edit to Display Datum D Dimension Symbol **d11** *(your "d" value may be different)*

214

Click: **d= Relations** from the Tools Ribbon > **▾ Local Parameters** to see the parameters of the part > *type (or pick from the model)* **d11=d6/2** in the Relations field *(your "d" values may be different)* > ☑ **Execute/Verify Relations** [Fig. 5.18(e)] > **OK** from the Verify Relations dialog box > **OK** > **Switch Symbols** to display the dimension values > **Ctrl+R** > **Ctrl+D** > **Ctrl+S** > **OK** > **LMB** to deselect

**Figure 5.18(e)** Relations **d11=d6/2** *(your **d** values may be different)*

The slot will now remain in the center of the upper protrusion regardless of changes in the protrusions width. In the real world, you will seldom encounter a situation where the project is designed and modeled without a "design change" or **ECO** (Engineering Change Order). Therefore, let us assume that an ECO has been "issued" that states: *the location of the hole on the angled surface must be aligned with the center of the slot at all times.* A relation could be established to control the hole's position as was done with the slot; but instead of referencing the hole from Datum B with a dimension, we will *align* the holes offset reference from the D datum.

Select the large hole in the Model Tree or on the model > press **RMB** > **Edit Definition** [Fig. 5.18(f)] > **Placement** tab > pick on datum **B (DATUM)** in the Offset References collector > **RMB** > **Remove** [Fig.

5.18(g)]

**Figure 5.18(f)** Edit Definition

**Figure 5.18(g)** Remove Offset Reference B (Datum)

Press and hold down the **Ctrl** key and select **D (DATUM)** from the model or Model Tree [Fig. 5.18(h)] > release the **Ctrl** key > Offset References collector **Offset**, next to D (to activate) > **Offset** > **Align** [Fig. 5.18(i)] > **MMB** > in the Graphics Window, **LMB** to deselect > **Ctrl+S** > **Enter** > **Ctrl+D**

**Figure 5.18(h)** Select D (Datum)

**Figure 5.18(i)** Align (Keep **2.0625** as the Offset distance from the edge for the first Offset Reference)

relations

> **Relations** and Parameters
> **Relations** and Parameters(1)
> d= **Relations**...
> d= **Relations**...(1)

## Relations and Parameters : ANCHOR

**PART NAME** : ANCHOR

### Relation Table

| Relation | Parameter | New Value |
|----------|-----------|-----------|
| | **Relations for ANCHOR:** | |
| | **Initial Relations** | |
| $d11 = d6 / 2$ | D11 | 9.375000e-01 |

### Local Parameters

| Symbolic constant | Current value | TYPE | SOURCE | ACCESS | DESIGNATE |
|-------------------|---------------|------|--------|--------|-----------|
| DESCRIPTION | --- | String | User-Defined | Full | YES |
| MODELED_BY | --- | String | User-Defined | Full | YES |

### Alternative Mass Property Parameters

| Symbolic constant | Current value | TYPE | SOURCE | ACCESS |
|-------------------|---------------|------|--------|--------|
| PRO_MP_SOURCE | GEOMETRY | String | Alternate Mass Properties | Full |
| MP_DENSITY | 2.827713e-01 | Real Number | Alternate Mass Properties | Full |
| PRO_MP_ALT_MASS | --- | Unknown | Alternate Mass Properties | Full |
| PRO_MP_ALT_VOLUME | --- | Unknown | Alternate Mass Properties | Full |
| PRO_MP_ALT_AREA | --- | Unknown | Alternate Mass Properties | Full |
| PRO_MP_ALT_COGX | --- | Unknown | Alternate Mass Properties | Full |
| PRO_MP_ALT_COGY | --- | Unknown | Alternate Mass Properties | Full |

**Figure 5.18(j)** Relations and Parameters in the Browser

Select **d6**, and **d11** (your **d** values may be different) in the Browser [Fig. 5.18(k)] to display them in the Graphics Window > [x] or [⊘] to close the Browser > in the Graphics Window, **LMB** > **Ctrl+R** > **Ctrl+D** > **Ctrl+S** > **Enter**

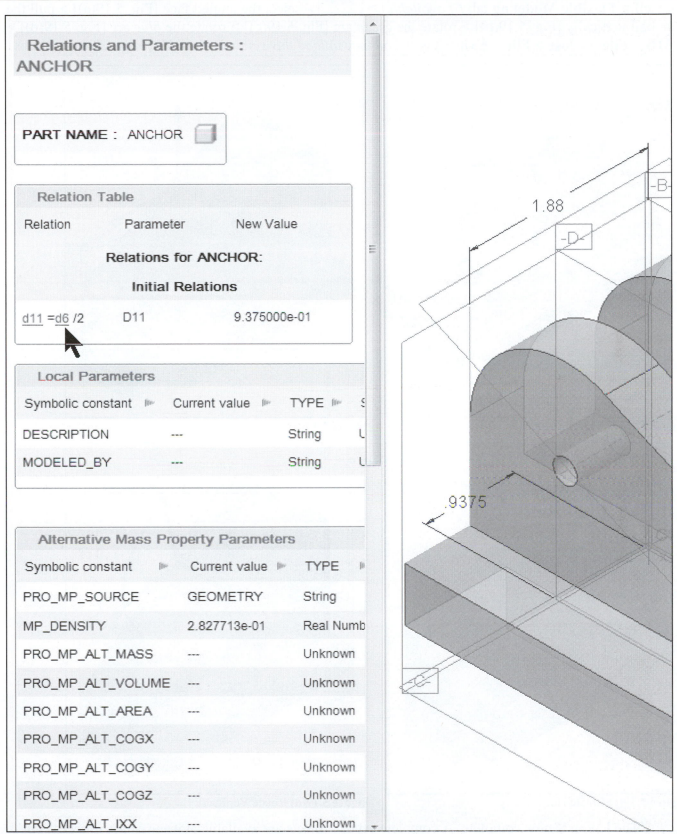

**Figure 5.18(k)** Displays Relation

Click: **File > Manage File > Delete Old Versions > Enter > File > Save As >** Type  **> Zip File (*.zip) > OK >** upload to your course interface or attach to an email and send to your instructor and/or yourself **> Flexible Modeling** tab *(if available) >* **>** select the angled face [Fig. 5.19(a)] > pull the *blue* ball vertically [Fig. 5.19(b)] > rotate the *green* arc [Fig. 5.19(c)] > rotate the *blue* arc [Fig. 5.19(d)] > **MMB > File > Close > File > Exit > Yes >** *You may upload this version of the file for grading*

**Figure 5.19(a)** Flexible Modeling

**Figure 5.19(b)** Pull

**Figure 5.19(c)** Rotate Horizontally

**Figure 5.19(d)** Rotate Vertically

The lesson is now complete. A complete set of Projects are available at ***www.cad-resources.com.***

# Lesson 6 Revolved Features

**Figure 6.1(a)** Clamp Foot

**Figure 6.1(b)** Clamp Ball

**Figure 6.1(c)** Clamp Swivel

## OBJECTIVES

- Master the **Revolve Tool** [Figs. 6.1(a-c)]
- Create **Chamfers** along part edges
- Understand and use the **Navigation browser**
- Alter and set the **Items** and **Columns** displayed in the **Model Tree**
- Create standard **Tapped Holes**
- Create **Cosmetic Threads** and complete **tabular information** for threads
- Edit **Dimension Properties**
- Use the **Model Player** to extract information and dimensions
- Get a hard copy using the **Print** command

## REFERENCES AND RESOURCES

For **Resources** go to **www.cad-resources.com** > click on the Creo Parametric 2.0 Book cover

- Lesson 6 Lecture
- Lesson 6 3D models embedded in a PDF
- Book Projects PDF
- Project Lectures

# REVOLVED FEATURES

The **Revolve Tool** creates a *revolved solid* or a *revolved cut* by revolving a sketched section around a centerline from the sketching plane [Fig. 6.2(a)]. You can have any number of centerlines in your sketch/section, but only one will be used to rotate your section geometry. Rules for sketching a revolved feature include:

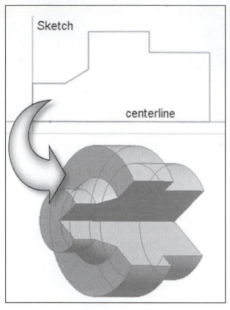

**Figure 6.2(a)** Revolved Section

- A section to be revolved must reference an axis
- By default, Creo Parametric 2.0 allows only one *axis of revolution* per sketch (within a sketch, you may select from different construction centerlines to be designated as the one *axis of revolution*)
- The geometry must be sketched on only one side of the *axis of revolution*
- The section must be closed for a solid [Fig. 6.2(b)] but can be open for a cut or a protrusion with assigned thickness

A variety of geometric shapes and constructions are used on revolved features. For instance, **chamfers** are created at selected edges of the part. Chamfers are *pick-and-place* features.

**Threads** can be a *cosmetic feature* representing the *nominal diameter* or the *root diameter* of the thread. Information can be embedded in the feature. Threads show as a unique color. By putting cosmetic threads on a separate layer, you can hide and unhide them.

| | |
|---|---|
| Revolved Solid Protrusion- closed section | |
| Revolved Protrusion with an assigned thickness- closed section | |
| Revolved Protrusion with an assigned thickness- open section | |
| Revolved Cut- open section | |
| Revolved Surface- open section | |

**Figure 6.2(b)** Types of Revolved Features

# Chamfers

Chamfers are created between abutting edges of two surfaces on the solid model. An edge chamfer removes a flat section of material from a selected edge to create a beveled surface between the two original surfaces common to that edge. Multiple edges can be selected.

There are four basic dimensioning schemes for edge chamfers (Fig. 6.3):

**Figure 6.3** Chamfer Options

- **45 x d** Creates a chamfer that is at an angle of **45°** to both surfaces and a distance **d** from the edge along each surface. The distance is the only dimension to appear when edited. **45 x d** chamfers can be created only on an edge formed by the intersection of two *perpendicular* surfaces.
- **d x d** Creates a chamfer that is a distance **d** from the edge along each surface. The distance is the only dimension to appear when edited.
- **d1 x d2** Creates a chamfer at a distance **d1** from the selected edge along one surface and a distance **d2** from the selected edge along the other surface. Both distances appear along their respective surfaces.
- **Ang x d** Creates a chamfer at a distance **d** from the selected edge along one adjacent surface at an **Angle** to that surface.

## Threads

Cosmetic threads are displayed with *magenta/purple* lines and circles. Cosmetic threads can be external or internal, blind or through. A cosmetic thread has a set of embedded parameters that can be defined at its creation or later, when the thread is added.

## Standard Holes

Standard holes are a combination of sketched and extruded geometry. It is based on industry-standard fastener tables. You can calculate either the tapped or clearance diameter appropriate to the selected fastener. You can use Creo Parametric 2.0 supplied standard lookup tables for these diameters or create your own. Besides threads, standard holes can be created with chamfers.

## Navigation Window

Besides using the File command and corresponding options, the *Navigation window* can be used to directly access other functions.

As previously mentioned, the working directory is a directory that you set up to contain Creo Parametric 2.0 files. You must have read/write access to this directory. You usually start Creo Parametric 2.0 from your working directory. A new working directory setting is not saved when you exit the current Creo Parametric 2.0 session. By default, if you retrieve a file from a non-working directory, rename the file and then save it, the renamed file is saved to the directory from which it was originally retrieved, if you have read/write access to that directory.

The navigation area is located on the left side (default) of the Creo Parametric 2.0 main window. It

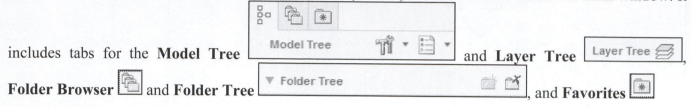

includes tabs for the **Model Tree** and **Layer Tree** ,

**Folder Browser** and **Folder Tree** , and **Favorites**

## Folder Browser

The Folder browser ( **Folder Browser**) (Fig. 6.4) is an expandable tree that lets you browse the file systems and other locations that are accessible from your computer. As you navigate the folders, the contents of the selected folder appear in the Creo Parametric 2.0 browser as a Contents page. The Folder browser contains top-level nodes for accessing file systems and other locations that are known to Creo Parametric 2.0:

**In Session** Creo Parametric 2.0 objects that have been retrieved into local memory.

**Desktop** Files and programs on the Desktop. Only Creo Parametric 2.0 items can be opened.

**My Documents** Files created and saved in Windows *My Documents* folder.

**Working Directory** Directory linked with the Select Working Directory command. All work will be accessed and saved in this location.

**Figure 6.4** Folder Browser

**Network Neighborhood** *(only for Windows)* The navigator shows computers on the networks to which you have access.

**Manikin Library** A Manikin model is a standard assembly that can be manipulated and positioned within the design scenario.

**Favorites** Saved folder locations for fast retrieval.

## Manipulating Folders

To work with folders, you can use the **Browser** to accomplish most file management requirements. The Browser's **Views** drop-down menu includes a (simple) **List, Thumbnails,** and **Details**:

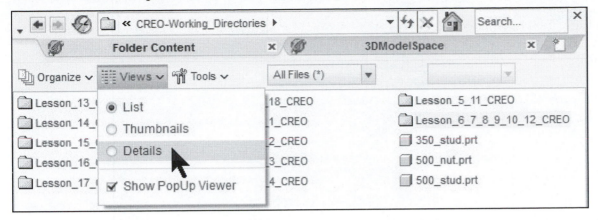

You can perform a variety of tasks from the Browser's **Organize** drop-down menu including **New Folder, Rename, Cut, Copy, Paste, Delete,** and **Add to common folders**:

The Browser's **Tools** drop-down menu includes **Sort By,** go **Up One Level, Add to Favorites, Remove from Favorites, Organize Favorites,** show **All Versions, Show Instances,** and **Send to Mail Recipient (as Zipped Attachment)**:

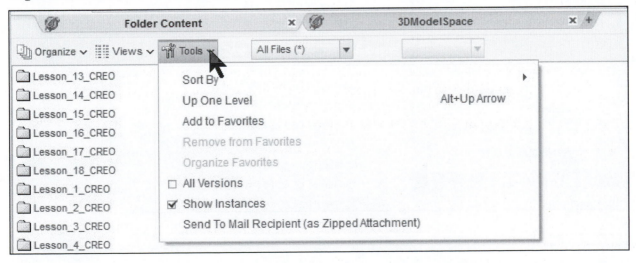

*Do not change your working directory from your default system work folder unless instructed to do so.*

From the Browser's **Type** field, you can limit the search to one of Creo Parametric 2.0's file types:

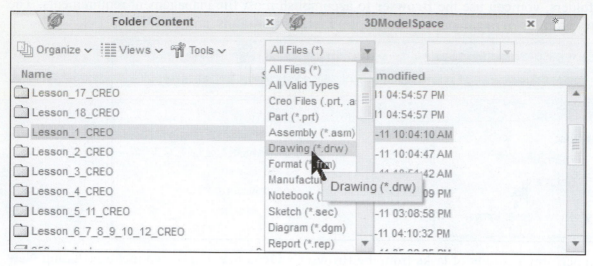

To set a working directory using the Browser, pick on the desired folder > ***RMB > Set Working Directory***:

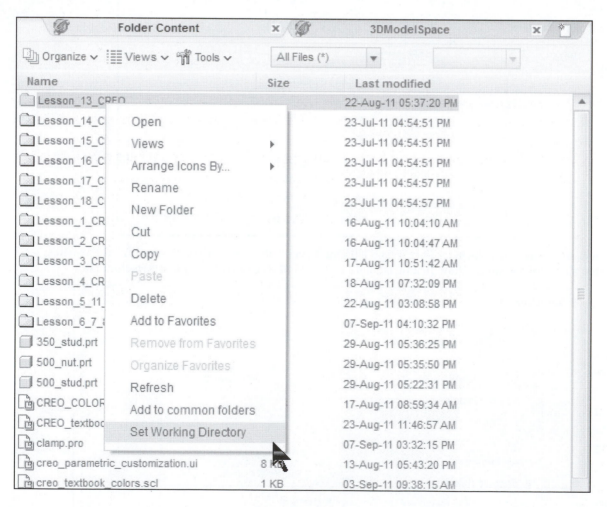

After you set (select) the working directory, pick on the Creo Parametric 2.0 object that you wish to preview:

Within the preview window, dynamically reorient the model using the **MMB** by itself (**Spin**- press and hold the **MMB** as you move the mouse) or in conjunction with the **Shift** key (**Pan**) or **Ctrl** key (**Zoom, Turn**). A mouse with a thumbwheel (as the middle button) can be rolled to zoom in/out on the object. Double-clicking on the file name will open the part. After you complete this lesson you will have this same part in your working directory.

From the Browser's toolbar you can also navigate to other directories or computer areas:

*Do not change your working directory from your default work folder unless you are instructed to do so by your teacher or you are on your own computer.*

**Figure 6.5** Clamp Foot Detail

## Clamp Foot

The Clamp Foot (Fig. 6.5) is the first of three revolved parts to be created in this Lesson. The Clamp Foot, Clamp Ball, and Clamp Swivel are revolved parts needed for the Clamp assembly and drawings later in the text. This part requires a revolved extrusion (protrusion), a revolved cut, a chamfer and rounds.

Launch **Creo Parametric 2.0 > Select Working Directory** from the Home Ribbon > navigate to your directory > **OK > Ctrl+N >** *type* **clamp_foot > OK > File > Prepare > Model Properties >** Units change > **Inch lbm Second (Creo Parametric Default) > Close > Close > File > Options > Configuration Editor > Import/Export > Import configuration file >** select **CREO_textbook.pro** [Fig. 6.6(a)] **> Open**

**Figure 6.6(a)** File Open Dialog Box

Click: **Display Filters** [Fig. 6.6(b)] >

☐ Show current session options
☐ Show options located in Favorites

*(uncheck both)* > Slide the vertical scroll bar *(right side of Creo Parametric Options dialog box)* down to the option *default_dec_places*. To the right of *default_dec_places*, click in the Value field > *type* **3** [Fig. 6.6(c)] > **Enter** *(status* ✳ *changed)*

**Figure 6.6(b)** default_dec_places

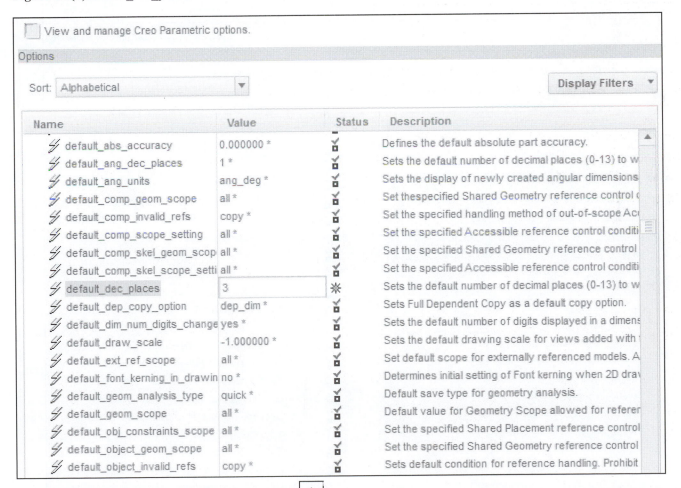

**Figure 6.6(c)** default_dec_places   Value: 3   Status: ✳

Click: **Add ...** > in the Option name field, *type*: **def_layer** *(you must type this with the keyboard, **but do not press Enter)** [Fig. 6.6(d)] > Option value:  > **layer_axis** > move the mouse pointer to the right side of the text: layer_axis, click in the Option value field > type *(space)* **datum_axes** > **OK** [Fig. 6.6(e)]

**Figure 6.6(d)** Option: def_layer layer_axis datum_axes

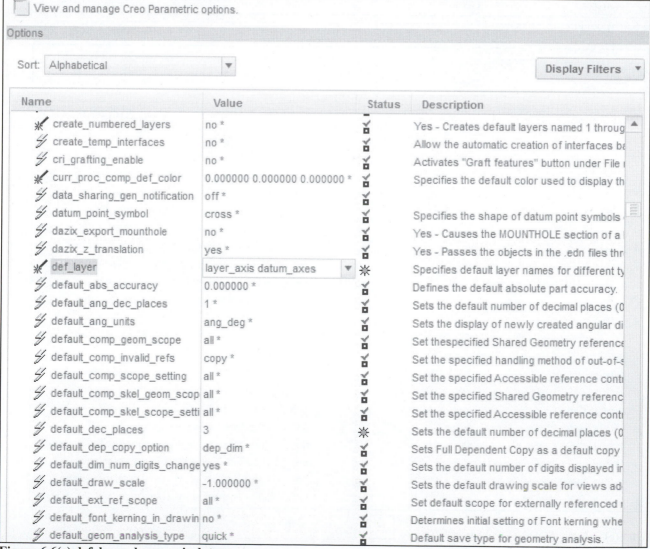

**Figure 6.6(e) def_layer, layer_axis datum_axes**

Click: **Display Filters** >  [Fig. 6.6(f)] > **OK** [Fig. 6.6(g)] > **Yes** > File name: *type* **clamp.pro** [Fig. 6.6(h)] > **OK** > **Ctrl+S** > **Enter**

**Figure 6.6(f)** Showing current session options with def_layer selected *(your selected/highlighted name may be different)*

**Figure 6.6(g)** Creo Parametric Options Confirmation Dialog Box

**Figure 6.6(h) clamp.pro**

Click: **View** tab >  on > on > **Model** tab > ⌀ Revolve > in the
Graphics Window, press **RMB** > **Define Internal Sketch** > Sketch Plane--- Plane: select datum **RIGHT**
[Fig. 6.7(a)] > Orientation **Left** > **Sketch** > **Sketch View** [Fig. 6.7(b)] *(To change the Graphics Toolbar location > RMB anywhere inside the Graphics Toolbar > Location > select an option)*

**Figure 6.7(a)** Sketch Plane and TOP:F2(DATUM PLANE) Reference

**Figure 6.7(b)** from the Graphics Toolbar

In the Graphics Window, press: **RMB > Construction Centerline** sketch a *vertical* centerline through the default coordinate system [Fig. 6.7(c)] > move the mouse off of the centerline > **MMB** > press **RMB** > **Designate Axis of Revolution** [Fig. 6.7(d)] > press **RMB** > **Line Chain** > sketch six lines > **MMB** > **MMB** > **LMB** [Fig. 6.7(e)] *(if incorrect, delete the lines and start again after clicking:* ⟲ *Undo)*

**Figure 6.7(c)** RMB > Construction Centerline                    **Figure 6.7(d)** Designate Axis of Revolution

**Figure 6.7(e)** Sketch Six Lines to Form a Closed Section (on one side of the centerline) *(your values will be different)*

Select one of the *weak* dimensions (highlights) > press **RMB** > ☐ Round Display Value (uncheck) > press **RMB** > ☐ Round Display Value (check) > **LMB** to deselect > press **RMB** > **Dimension** > add the vertical dimension by picking both horizontal lines > **MMB** to place the dimension [Fig. 6.7(f)] > move the pointer > **MMB** > **MMB** > capture all of the sketch and dimensions with a window > press **RMB** > **Modify** > **Lock Scale** [Fig. 6.7(g)] > *type* **1.25** for the large diameter > **Enter** > ☑ [Fig. 6.7(h)]

**Figure 6.7(f)** Sketcher Dimensioning *(your values will be different)*

**Figure 6.7(g)** Modify Dimensions Dialog Box    **Figure 6.7(h) 1.25** Diameter

234

For each dimension, double-click on and change to the design value [Fig. 6.7(i)] > **LMB** to deselect > reposition the dimensions as shown > in the Graphics Window, press **RMB** > **OK** > in the Graphics Window, press **RMB** > **Show Section Dimensions** [Fig. 6.7(j)]

**Figure 6.7(i)** Design Value Dimensions

**Figure 6.7(j)** Show Section Dimensions

**Figure 6.7(k)** Revolve Feature Preview

**Figure 6.7(l)** Revolved Extrusion

Click: Revolve > **Remove Material** > **Placement** tab > **Define** > Use Previous [Figs. 6.8(a-b)] > **Sketch View** [Fig. 6.8(c)] > in the Graphics Window, press **RMB** > **Axis of Revolution** sketch a *vertical* axis through the default coordinate system [Fig. 6.8(d)] > move the mouse off of the axis > **MMB** > **LMB** > from the Graphics Toolbar > **Hidden Line** [Fig. 6.8(e)]

**Figure 6.8(a)** Use Previous

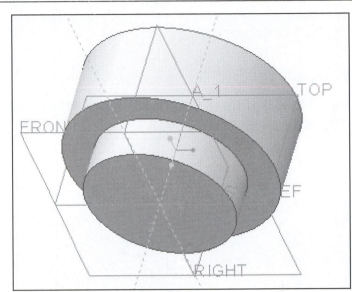

**Figure 6.8(b)** 3D Sketch View

**Figure 6.8(c)** Sketch View

**Figure 6.8(d)** Sketch a Vertical Axis of Revolution

**Figure 6.8(e)** Hidden Line Display

From the Sketch Ribbon, click: ⊡ next to ↻ ▾ > Center and Ends > sketch the arc *(LMB to place the center of the arc > LMB to place the start point of the arc > LMB to place the end point of the arc)* [Fig. 6.8(f)] > **MMB** > **LMB** [Fig. 6.8(g)] > press **RMB** > **Line Chain** sketch a *vertical* line ending at

☐ (Select All)
☐ Axis Display
☐ Point Display
☐ Csys Display
☐ Plane Display

datum FRONT [Figs. 6.8(h-i)] > **MMB** > **MMB** to end the current tool > ⊡ > > **LMB** in the Graphics Window > **LMB** to deselect

**Figure 6.8(f)** Sketch the Arc

**Figure 6.8(g)** Completed Arc

**Figure 6.8(h)** Sketch the Vertical Line

**Figure 6.8(i)** Completed Vertical Line

Click:  **Delete Segment** [Fig. 6.8(j)] > press and hold the **LMB** and draw a spline through the two elements you wish to remove [Fig. 6.8(k)] *(Note: if you delete the wrong items, click:* ↺ *Undo and try*

*again)* > release the **LMB** to complete the trim > **MMB** > [Fig. 6.8(l)] > **Close**

**Figure 6.8(j)** Dynamically Trim the Arc and Line

**Figure 6.8(k)** Trimmed Sketch

**Figure 6.8(l)** Feature Requirements

239

Click:  > **LMB** in the Graphics Window > press **RMB** > **Dimension** > *double-click* on the arc edge > **MMB** to place the diameter dimension > *type* **.450** [Fig. 6.8(m)] > **Enter** > add a

dimension from the center of the arc to the bottom of the part > *type* **.200** [Fig. 6.8(n)] > **Enter** > to end the current tool > double-click on the hole diameter dimension > *type* **.425** > **Enter** > move the mouse off of the dimension > **LMB** to deselect

**Figure 6.8(m)** Correct Dimensioning Scheme *(your dimension values **will** be different)*

**Figure 6.8(n)** Modify Dimensions to **.450**, **.200**, and **.425** (these are the design values- see Figure 6.5)

Click: **File > Save As > Save a Copy >** *(Saves the section sketch. The sketch will be used in the next component. If you do not save it here, it will not be available in your palette.)* >

**OK >**  [Fig. 6.8(o)] > **Ctrl+D >** [Fig. 6.8(p)] > [Fig. 6.8(q)] > **Ctrl+S** > **Enter >** in the Graphics Window, **LMB** to deselect

**Figure 6.8(o)** Previewed Revolved Cut

**Figure 6.8(p)** Revolved Cut Shading with Edges

**Figure 6.8(q)** Completed Cut

241

Press: **MMB** spin the model > pick on the revolved protrusion > pick on the edge of the cut until selected

*(highlights)* [Fig. 6.8(r)] > [Fig. 6.8(s)] > double-click on the dimension and modify to the design value of **.03125** [Fig. 6.8(t)] > **Enter** > **MMB** [Fig. 6.8(u)] > in the Graphics Window, **LMB** to deselect > ⊟ > **Enter**

**Figure 6.8(r)** Pick on Edge

**Figure 6.8(s)** Chamfer Preview

**Figure 6.8(t)** Chamfer Dimension

**Figure 6.8(u)** Completed Chamfer

Select one edge, then press and hold the **Ctrl** key and pick on the remaining three edges > release the **Ctrl** key > press **RMB** > **Round Edges** [Fig. 6.8(v)] > double-click on the dimension and modify to the design value of **.03125** [Fig. 6.8(w)] > **Enter** > **MMB** > **Ctrl+S** > **OK** > **LMB** to deselect > **Ctrl+D** > **Ctrl+R**

Figure 6.8(v) Round Edges

Figure 6.8(w) .03125 Round

From the Model Tree, pick: **FRONT** > **RMB** > **Properties** > Name- *type* **A** > [   A   ] > [ ● On Datum ] > **OK** (from the Datum dialog box) [   ] > pick **RIGHT** > **RMB** > **Properties** > Name- *type* **B** > [   A   ] > [ ● On Datum ] > **OK** > pick **TOP** > **RMB** > **Properties** > Name- *type* **C** > [   A   ] > [ ● On Datum ] >**OK** > slowly pick twice on the coordinate system > *type* [ CLAMP_FOOT_CSYS ] > **Enter** [Fig. 6.8(x)] > **File** > **Prepare** > **Model Properties** > Material **change** > [ steel.mtl ] > [ ▶▶▶ ] > **OK** > **Close** > **LMB** > **Ctrl+S** > **OK** > **File** > **Save As** > **Save a Copy** > Type [ ▼ ] > **Zip File (*.zip)** > **OK** > [   ] **Close**

Figure 6.8(x) Completed Clamp Foot

**Figure 6.9** Clamp Ball Detail

## Clamp Ball

The second part for this lesson is the Clamp Ball (Fig. 6.9). Much of the sketching is the same for this part as was done for the first revolved feature of the Clamp Foot. Instead of an internal revolved cut, you will create a standard hole. The Clamp Ball is made of *nylon*.

---

Click: ⬜ **New** > Name-- *type* **clamp_ball** > **OK** > **File** > **Options** > **Configuration Editor** > **Import/Export** > **Import configuration file** > **clamp.pro** [Fig. 6.10(a)] from your working directory > **Open** > **OK** > **No** > **File** > **Prepare** > **Model Properties** > Units **change** > **Inch lbm Second (Creo Parametric Default)** > **Close** > Material **change** > 🗎 nylon.mtl > ▶▶▶ [Fig. 6.10(b)] > **OK** > **Close**

---

**Figure 6.10(a)** Configuration File

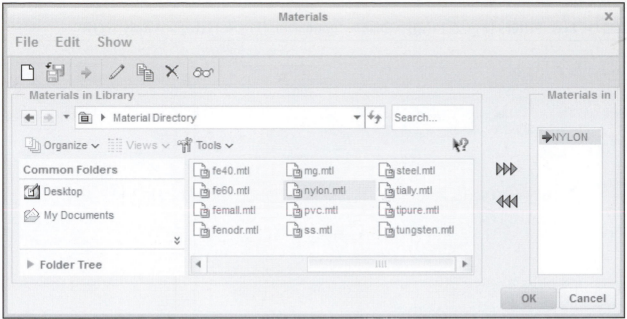

**Figure 6.10(b)** NYLON Material

## Model Tree

The Model Tree is a tabbed feature on the Creo Parametric 2.0 navigator that contains a list of every feature (or component) in the current part, assembly, or drawing. The model structure is displayed in hierarchical (tree) format with the root object (the current feature, part or assembly) at the top of its tree and the subordinate objects (features, parts, or assemblies) below.

If you have multiple Creo Parametric 2.0 windows open, the Model Tree contents reflect the file in the current "active" window. The Model Tree lists only the related feature- and part-level objects in a current file and does not list the entities (such as edges, surfaces, curves, and so forth) that comprise the features.

Each Model Tree item displays an icon that reflects its object type, for example, assembly, part, feature, or datum plane (also a feature). The icon can also show the display status for a feature, part, or assembly, for example, suppressed or hidden.

You can save the Model Tree as a .txt file. Selection in the Model Tree is object-action oriented; you select objects in the Model Tree without first specifying what you intend to do with them. Items can be added or removed from the Model Tree column display using Settings in the Navigator:

- Select objects, and perform object-specific operations on them using the shortcut menu.
- Filter the display by item type, for example, hiding or un-hiding datum features.
- Open a part within an assembly file by right-clicking the part in the Model Tree.
- Create or modify features and perform other operations such as deleting and redefining.
- Search the Model Tree for model properties or other feature information.
- Show the display status for an object, for example, suppressed or hidden.

Click:  > **Tree Filters** [Fig. 6.11(a)] > Display options *on* [Fig. 6.11(b)] > **Apply** > **OK**

Figure 6.11(a) Tree Filters

Figure 6.11(b) Model Tree Items Dialog Box

Click:  in the Navigator > **Tree Columns** [Fig. 6.12(a)] (the Model Tree Columns dialog opens)

Figure 6.12(a) Model Tree Columns

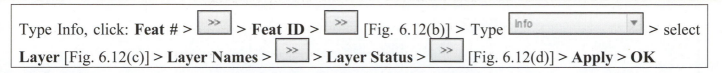
Type Info, click: **Feat #** > [>>] > **Feat ID** > [>>] [Fig. 6.12(b)] > Type [Info ▼] > select **Layer** [Fig. 6.12(c)] > **Layer Names** > [>>] > **Layer Status** > [>>] [Fig. 6.12(d)] > **Apply** > **OK**

**Figure 6.12(b)** Model Tree Columns Dialog Box, Type: Info

**Figure 6.12(c)** Model Tree Columns Dialog Box, Type: Layer

Figure 6.12(d) Model Tree Columns Dialog Box, Displayed List

Press: **LMB** on the sash [⊪] and drag to expand the Model Tree [Fig. 6.12(e)] > release the **LMB** > adjust the width of each column by dragging the column divider Feat ID✛Layer Names > [⊪] drag the sash to the left to decrease the Model Tree size [Fig. 6.12(f)]

Figure 6.12(e) Expand the Model Tree and Adjust the Width of the Columns

Figure 6.12(f) Decrease the Model Tree Width

Click: 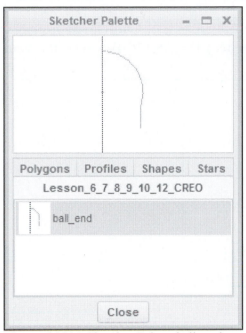 *off* > pick on datum **RIGHT** in the Graphics Window > | Revolve | > | ⇄ | > | Palette | > select

the tab with your working directory name [Fig. 6.13(a)] > double-click | ball_end | > **LMB** to place the sketch [Fig. 6.13(b)], dashboard displays [Fig. 6.13(c)] and the sketch is initially is placed [Fig. 6.13(d)] *(If you do not have a ball_end sketch, you did not save it when sketching the ball end cut in the Clamp_ball. You must edit definition the section in the Clamp_ball or sketch the shape from scratch.)*

**Figure 6.13(a)** Sketcher Palette ball_end

**Figure 6.13(b)** Place the sketch with your LMB

**Figure 6.13(c)** Dashboard

**Figure 6.13(d)** Initially Placed ball_end Sketch

With the pointer on the move handle ⊗, press and hold the **RMB**, move the pointer to the upper arc end of the ball_end sketch [Fig. 6.13(e)] > release the **RMB** > with the pointer on the move handle, press **LMB** and move the sketch to the *vertical* (TOP) reference [Fig. 6.13(f)] > release the **LMB** > in the Dashboard, Scale **2** > [2.000000] > **Enter** > ✓ > double-click on the long vertical dimension [228.4117591214] > *type* **.75** [.75 12] > **Enter** > **LMB** to deselect > **Close** the Sketcher Palette > ✕ > □ Csys Display > **LMB** in the Graphics Window [Fig. 6.13(g)] > ⊙ **Coincident** from the Constrain Group in the Sketch tab Ribbon > pick the **Line (Reference)** [Fig. 6.13(h)] > pick the **End Point 1: Line** [Fig. 6.13(i)] > **MMB** [Fig. 6.13(j)]

**Figure 6.13(e)** Repositioned Handle

**Figure 6.13(f)** Move Sketch

**Figure 6.13(g)** Placed Sketch

**Figure 6.13(h)** Select the Line Reference

**Figure 6.13(i)** Select the End Point

**Figure 6.13(j)** Coincident

Press: **RMB** > **Line Chain** sketch the two lines to close the section [Fig. 6.13(k)] > **MMB** > **MMB** > **LMB** > double-click on each dimension and modify to the design values (Fig. 6.9) of **1.500** diameter, **.875** diameter, and **.750** > **LMB** to deselect > reposition the dimensions [Fig. 6.13(l)] > ☑ > **Ctrl+D** [Fig. 6.13(m)] > [toolbar icons: □ ▢ ↻ | InternalCL | ⊥▾ | 360.00 ▾ | ⅔ ▨ ⊏ ] > ☑ > 🔍 > ⊹ > drag to expand the Model Tree window (to the right) to see the columns > double-click ⊹ to shut the Model Tree window > in the Graphics Window, **LMB** to deselect > **Ctrl+S** > **Enter** [Fig. 6.13(n)]

**Figure 6.13(k)** Closed Sketch

**Figure 6.13(l)** Modified and Repositioned Dimensions

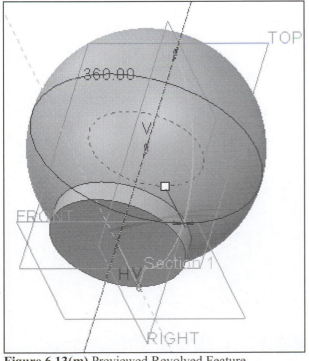

**Figure 6.13(m)** Previewed Revolved Feature

**Figure 6.13(n)** Completed Feature

## Standard Holes

*Hole charts* are used to lookup diameters for a given fastener size. You can create custom hole charts and specify their directory location with the configuration file option *hole_parameter_file_path*. UNC, UNF and ISO hole charts are supplied with Creo Parametric 2.0. Create a standard **.500-13 UNC-2B** hole, **.375** thread depth, **.50** tap drill, and a standard chamfer.

Click: [icon] *on* > [Hole] > [icon] > [1/2-13] > [icon] **Adds countersink** > **Shape** tab > [Variable 0.375000] [Fig. 6.14(a)] > countersink (.03125*2 + .4219 = **.4844**) > angle **90.0** > **.500** depth [icon] [icon] [icon] [icon] [UNC] [1/2-13] [0.5000] [icon] [icon] [icon] > **View** tab > **MMB** spin the part > [icons] *on* > **Hole** tab > **Placement** tab > pick on axis **A_1** *(your axis name may be different)* > press and hold the **Ctrl** key and pick on the flat surface [Fig. 6.14(b)]

**Figure 6.14(a)** Standard Hole Shape

**Figure 6.14(b)** Placement Tab *(your axis name may be different)*

252

Release the **Ctrl** key, click: **Note** tab [Fig. 6.14(c)] > 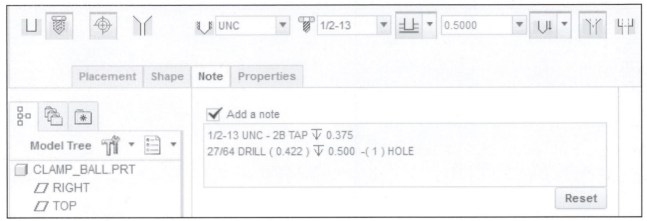 to open the Model Tree *(lower left-hand corner)* > click ▶ next to Hole 1 in the Model Tree > **Note_0** > **RMB** > **Properties** Note dialog box opens [Figs. 6.15(a-b)]

**Figure 6.14(c)** Hole Note

**Figure 6.15(a)** Note Properties

**Figure 6.15(b)** Note Dialog Box

In the Note Dialog Box, modify the Name and Text fields [Fig. 6.15(c)]. Move the mouse pointer within the Text field, click: **LMB > Symbols >** 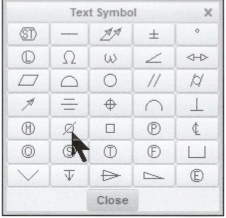 add the diameter symbol [Fig. 6.15(d)] > **OK** from the Note dialog box > **MMB** spin the model [Fig. 6.15(e)] > **Ctrl+D > Ctrl+S > OK**

**Figure 6.15(c)** Modified Note

**Figure 6.15(d)** Text Symbol Dialog Box

**Figure 6.15(e)** Annotation

254

From the Model Tree, pick: **FRONT > RMB > Properties** > Name- *type* **A** >  > ⦿ On Datum > **OK** > pick **RIGHT > RMB > Properties** > Name- *type* **B** > > ⦿ On Datum > **OK** > pick **TOP > RMB > Properties** > Name- *type* **C** > > ⦿ On Datum > **OK** > slowly pick twice on the coordinate system in the Model Tree > *type* **CLAMP_BALL_CSYS > Enter** > in the Graphics Window, **LMB** to deselect > click ▶ next to Annotations in the Model Tree [Fig. 6.15(f)] > > **OK**

**Figure 6.15(f)** Set Datum Tag Annotations

Double-click on the model to display the dimensions for the protrusion > pick on the ∅.875 dimension > press **RMB** [Fig. 6.16(a)] > **Properties** Dimension Properties dialog box displays [Fig. 6.16(b)]

**Figure 6.16(a)** Dimension Displayed

**Figure 6.16(b)** Dimension Properties Dialog Box, Properties tab

The **Display** tab [Fig. 6.16(c)] shows the parametric dimension symbol. The **Text Style** tab [Fig. 6.16(d)] provides options for Character, and Note/Dimension variations.

**Figure 6.16(c)** Display Tab

**Figure 6.16(d)** Dimension Properties, Text Style Tab

Click: **Move** > select a new position for the ∅**.875** dimension [Fig. 6.16(e)] > **OK** > repeat to move the other dimensions as required *(pick on a dimension > press **RMB** > **Properties** > **Move** > select a new position > **OK**)* > **LMB** to deselect the dimension > move the pointer > **LMB** to deselect the protrusion

**Figure 6.16(e)** Moved Dimension

Pick on the model > pick on the edge [Fig. 6.17(a)] > press **RMB** > **Round Edges** > double-click on the dimension > *type* **.06125** [Fig. 6.17(b)] > **Enter** > **Enter** > in the Graphics Window, **LMB** to deselect

**Figure 6.17(a)** Select the Edge

**Figure 6.17(b)** Design Value Radius **.06125**

Click: 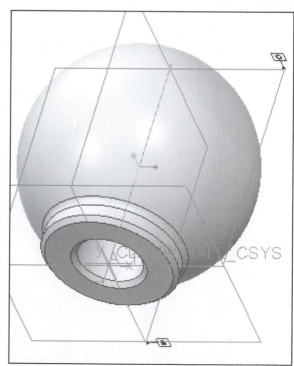 Chamfer > pick on the edge [Fig. 6.18(a)] > D x D  D 0.03125 > **Enter** > ✓ > **Ctrl+D** [Fig. 6.18(b)] > **Ctrl+S** > **Enter** > in the Graphics Window, **LMB** to deselect > **File** > **Save As** > **Save a Copy** > Type > **Zip File (*.zip)** > **OK** > **File** > **Close** > upload to your class interface and email to your instructor and yourself

**Figure 6.18(a)** Chamfer **.03125**

**Figure 6.18(b)** Completed Chamfer

**Figure 6.19** Clamp Swivel Detail

## Clamp Swivel

The Clamp Swivel is the third part created by revolving one section about a centerline (Fig. 6.19). The Clamp Swivel is a component of the Clamp Assembly (Fig. 6.20).

Click: [icon] **Create a new model** > Name-- *type* **clamp_swivel** > **OK**

*(If you have just started a new session of Creo Parametric 2.0, remember to load your configuration file >* ***File*** *>* ***Options*** *>* ***Configuration Editor*** *>* ***Import/Export*** *>* ***Import configuration file*** *> click on* ***clamp.pro*** *which was previously created and saved >* ***Open*** *>* ***OK*** *>* ***No)***

> **File** > **Prepare** > **Model Properties** > Units **change** *(check the unit setting* Inch lbm Second*)* > **Close**

**Figure 6.20** Clamp Swivel Highlighted in Clamp Assembly

Click: Material **change** >  **Create new material** [Fig. 6.21(a)] > in Name field, *type* **STEEL_1066** > fill in the form [Fig. 6.21(b)] > **Save To Library…** > navigate to your working directory > **OK**

**Figure 6.21(a)** Materials Dialog Box

**Figure 6.21(b)** Material Definition Dialog Box

Navigate to your working directory (select **Working Directory** from the Common Folders column) > select **steel_1066.mtl** >  [Fig. 6.21(c)] > **OK** to close the Materials dialog box > **Close** to close the Model Properties dialog box [Fig. 6.21(d)]

**Figure 6.21(c)** STEEL_1066

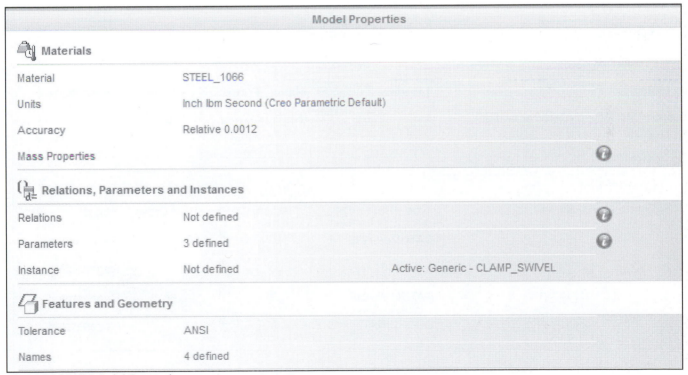

**Figure 6.21(d)** Model Properties

Rename the coordinate system to **CLAMP_SWIVEL_CSYS > Enter >** in the Graphics Window, **LMB** to deselect > 🖫 > **OK > Tools** tab > **Model** [Fig. 6.21(e)] > 🗗 to close the Browser [Fig. 6.21(f)]

**Figure 6.21(e)** Model Information

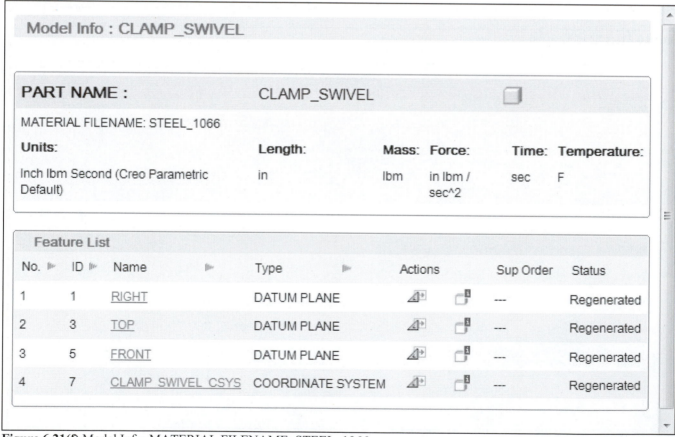

## Model Info : CLAMP_SWIVEL

| PART NAME : | CLAMP_SWIVEL | |
|---|---|---|

MATERIAL FILENAME: STEEL_1066

| Units: | Length: | Mass: | Force: | Time: | Temperature: |
|---|---|---|---|---|---|
| Inch lbm Second (Creo Parametric Default) | in | lbm | in lbm / sec^2 | sec | F |

### Feature List

| No. | ID | Name | Type | Actions | Sup Order | Status |
|---|---|---|---|---|---|---|
| 1 | 1 | RIGHT | DATUM PLANE | 🔺 🗗 | --- | Regenerated |
| 2 | 3 | TOP | DATUM PLANE | 🔺 🗗 | --- | Regenerated |
| 3 | 5 | FRONT | DATUM PLANE | 🔺 🗗 | --- | Regenerated |
| 4 | 7 | CLAMP_SWIVEL_CSYS | COORDINATE SYSTEM | 🔺 🗗 | --- | Regenerated |

**Figure 6.21(f)** Model Info: MATERIAL FILENAME: STEEL_1066

Click: **File** > **Options** > **Sketcher** > ☑ Show the grid ☑ Snap to grid > **OK** > **No** > **Model** tab > pick on the **RIGHT** datum plane > ⟨⟩ **Revolve** > ⟨⟩ > in the Graphics Window, press **RMB** > **Axis of Revolution** sketch a *vertical* axis through the default coordinate system > move the mouse off of the axis > **MMB** > **LMB** > ⟨⟩ > sketch the arc center and end points > **MMB** > **LMB** > press **RMB** > **Line Chain** sketch the lines all at different lengths [Fig. 6.22(a)] > **MMB** > **MMB** > **LMB** > **File** > **Options** > **Sketcher** > ☐ Show the grid ☐ Snap to grid > **OK** > **No** > ⟨⟩ from the Graphics Toolbar > ☐ (Select All) > **LMB** in the Graphics Window > press **RMB** > **Dimension** > change the ball-end radius to a diameter dimension, double-click on the arc > **MMB** to place the diameter value > move the mouse off of the dimension > **MMB** > re-dimension as shown > **MMB** > modify values as per design requirements [Fig. 6.22(b)] > ☑ > ☑

**Figure 6.22(a)** Sketch (your dimensions may be different)

**Figure 6.22(b)** Dimension, Modify and Reposition

263

Press: **Ctrl+D** >  > **OK** > from the Graphics Toolbar > [Select All] > **LMB** > datum **RIGHT** from the Model Tree > **RMB** > **Properties** > Name- *type* **B** > > On Datum > **MMB** > pick **FRONT** > **RMB** > **Properties** > Name- *type* **A** > > On Datum > **MMB** > pick **TOP** > **RMB** > **Properties** > Name- *type* **C** > > On Datum > **MMB** > **LMB** to deselect > **File** > **Options** > **Model Display** > **Trimetric** > **Isometric** > **OK** > **No** > **MMB** spin the model (Fig. 6.23)

**Figure 6.23** Completed Revolved Extrusion

Click: 冚 Hole > pick on datum **C** > press **RMB** > **Offset References Collector** > **Placement** tab > Flip > ∅ 0.500 > **Enter** > > > pick datum **A** > Offset **.5625** > **Enter** > press and hold the **Ctrl** key and pick on datum **B** > release the **Ctrl** key > **Offset** > **Offset** > **Align** [Fig. 6.24(a)]

**Figure 6.24(a)** Placement Tab (may have to use -0.5625 for A Datum Offset value)

Click: **Shape** tab > Side 2: **None** > **Through All** [Fig. 6.24(b)] > ✓ > **LMB** in the Graphics Window > **Ctrl+S** > **OK** > **LMB** to select the model > **LMB** to select the edge (Fig. 6.25) > press **RMB** > **Round Edges** > hold down the **Ctrl** key and select the remaining edges > release the **Ctrl** key > double-click on the dimension > *type* **.100** > **Enter** > **Ctrl+D** > **Enter** > **LMB** to deselect > **File** > **Save** > **OK**

**Figure 6.24(b)** Shape Tab

**Figure 6.25** Rounds

## Cosmetic Threads

A **cosmetic thread** is a feature that "represents" the diameter of a thread without having to show the actual threaded feature. Since a threaded feature is memory intensive, using cosmetic threads can save an enormous amount of visual memory on your computer. It is displayed in a unique default color. **Internal cosmetic threads** are created automatically when holes are created using the Hole Tool (Standard-Tapped). In situations where the internal thread is unique and the Hole Tool cannot be used, internal cosmetic threads can be added to the hole. **External cosmetic threads** represent the *root diameter*. For threaded shafts you must create the external cosmetic threads. The following table lists the parameters that can be defined for a cosmetic thread. In this table, "pitch" is the distance between two threads.

| PARAMETER NAME | PARAMETER TYPE | PARAMETER DESCRIPTION |
|---|---|---|
| **MAJOR_DIAMETER** | Real Number | Thread major diameter |
| **THREADS_PER_INCH** | Real Number | Threads per inch (1/pitch) |
| **THREAD FORM** | String | Thread form |
| **CLASS** | String | Thread class |
| **PLACEMENT** | Character | Thread placement (A-external, B-internal) |
| **METRIC** | Yes No (True/False) | Thread is metric |

Click: [ Engineering ▼ ] Engineering Group drop down > **Cosmetic Thread** > pick the cylindrical surface [Fig. 6.26(a)]

**Figure 6.26(a)** Select the Thread Surface

Pick the thread start surface--the edge lip surface [Fig. 6.26(b)] > in the Dashboard [⊥⌐ ▼] [4.000000 ▼] >
Enter > [⌀ 0.4485 ▼] thread root diameter: **.4485** [Fig. 6.26(c)] > **Enter** > **Properties** tab [Fig. 6.26(d)]

**Figure 6.26(b)** Select the Thread Start Surface

**Figure 6.26(c)** Root Diameter **.4485**

**Figure 6.26(d)** Properties Tab

After creating the cosmetic thread, edit the thread panel. The thread size of an external thread must be changed to the nominal size from the root diameter defaulted on the thread table. The task was to create an external cosmetic thread (**.500-13 UNC-2A**) using the ∅**.500** surface [Fig. 6.26(a)]. The thread started at the "neck" and extended **4.00** along the swivel's shaft [Fig. 6.26(b)]. The thread parameter [Fig. 6.26(e)] shows the major diameter as **0.4485**.

Since you are cosmetically representing the *root diameter (.4485)* of the external thread on the model, the *thread diameter* is *smaller* than the *nominal (.500)* thread size.

---

Click in each table field and change the values [Fig. 6.26(f)] > THREADS_PER_INCH **13** > FORM **UNC** > CLASS **2** > ✓ > in the Graphics Window, **LMB** to deselect > **Ctrl+S** > **OK**

---

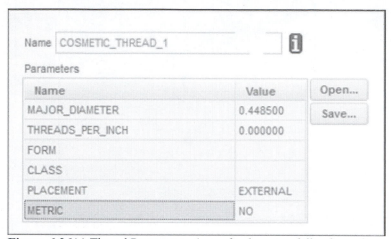

**Figure 6.26(e)** Thread Parameters *(your display may differ depending on the number of decimal places set)*

**Figure 6.26(f)** Edited Thread Parameters *(your display may differ depending on the number of decimal places set)*

268

Click: [Cosmetic Thread 1] in the Model Tree > **RMB** > **Edit Parameters** [Fig. 6.26(g)] > **OK** > **LMB** to deselect > [icon] > [Hidden Line] (to see the cosmetic surface) [Fig. 6.26(h)]

**Figure 6.26(g)** Thread Parameters

**Figure 6.26(h)** Cosmetic Thread

Select the cosmetic feature in the Model Tree > **RMB** > **Info** > **Feature** [Fig. 6.26(i)] > 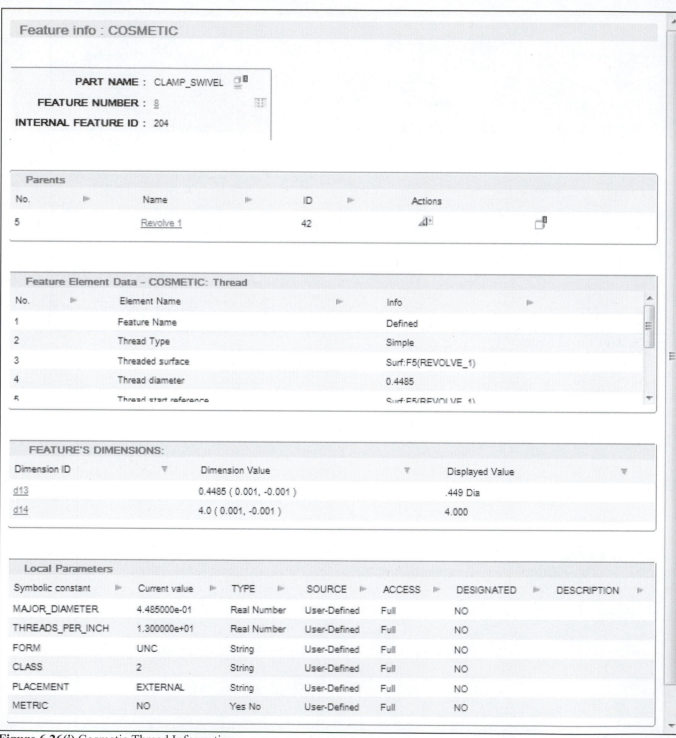 to close the Browser > **Ctrl+D** > in the Graphics Window, **LMB** to deselect > ⬜ > ⬛ **Shading With Edges** > 💾 > **OK** > **File** > **Manage File** > **Delete Old Versions** > **Enter** > **File** > **Save As** > **Save a Copy** > Type ▼ > **Zip File (\*.zip)** > **OK** > upload to your course interface or attach to an email and send to your instructor and/or yourself

**Feature info : COSMETIC**

|  | |
|---|---|
| PART NAME : | CLAMP_SWIVEL |
| FEATURE NUMBER : | 8 |
| INTERNAL FEATURE ID : | 204 |

**Parents**

| No. | Name | ID | Actions | |
|---|---|---|---|---|
| 5 | Revolve 1 | 42 | | |

**Feature Element Data – COSMETIC: Thread**

| No. | Element Name | Info |
|---|---|---|
| 1 | Feature Name | Defined |
| 2 | Thread Type | Simple |
| 3 | Threaded surface | Surf:F5(REVOLVE_1) |
| 4 | Thread diameter | 0.4485 |
| 5 | Thread start reference | Surf:F5(REVOLVE_1) |

**FEATURE'S DIMENSIONS:**

| Dimension ID | Dimension Value | Displayed Value |
|---|---|---|
| d13 | 0.4485 ( 0.001, -0.001 ) | .449 Dia |
| d14 | 4.0 ( 0.001, -0.001 ) | 4.000 |

**Local Parameters**

| Symbolic constant | Current value | TYPE | SOURCE | ACCESS | DESIGNATED | DESCRIPTION |
|---|---|---|---|---|---|---|
| MAJOR_DIAMETER | 4.485000e-01 | Real Number | User-Defined | Full | NO | |
| THREADS_PER_INCH | 1.300000e+01 | Real Number | User-Defined | Full | NO | |
| FORM | UNC | String | User-Defined | Full | NO | |
| CLASS | 2 | String | User-Defined | Full | NO | |
| PLACEMENT | EXTERNAL | String | User-Defined | Full | NO | |
| METRIC | NO | Yes No | User-Defined | Full | NO | |

**Figure 6.26(i)** Cosmetic Thread Information

270

# Using the Model Player

The **Model Player** (Fig. 6.27) lets you observe how a part is built. You can:

- Move backward or forward through the feature-creation history of the model.
- Regenerate each feature in sequence, starting from the specified feature
- Display each feature as it is regenerated or rolled forward
- Update (regenerate all the features in) the entire display when you reach the desired feature
- Obtain information about the current feature (you can show dimensions, obtain regular feature information, investigate geometry errors, and enter Fix Model mode)

You can select one of the following:

- **Regenerate features** Regenerates each feature in sequence, starting from the specified feature
- **Display each feature** Displays each feature in the graphics window as it is being regenerated
- **Compute CL** (Available in Manufacturing mode only) When selected, the CL data is recalculated for each NC sequence during regeneration

**Figure 6.27** Model Player

Select one of the following commands:

**Go to the beginning of the model** moves immediately to the beginning of the model

**Step backward through the model one feature at-a-time** and regenerates the preceding feature

**Stop play**

**Step forward through the model one feature at-a-time** and regenerates the next feature

**Go to the last feature in the model** moves immediately to the end of the model (resume all)

**Slider Bar** Drag the slider handle to the feature at which you want model playback to begin. The features are highlighted in the graphics window as you move through their position with the slider handle. The feature number and type are displayed in the selection panel [such as #4 (COORDINATE SYSTEM)], and the feature number is displayed in the **Feat #** box.

**Select feature from screen or model tree** lets you select a starting feature from the graphics window or the Model Tree. Opens the **SELECT FEAT** and **SELECT** menus. After you select a starting feature, its number and ID are displayed in the selection panel, and the feature number is displayed in the **Feat #** box.

Feat # 5    of 5   Let's you specify a starting feature by typing the feature number in the box. After you enter the feature number, the model immediately rolls or regenerates to that feature.

To stop playback, click the **Stop play** button. Use the following commands for information:

- **Show Dims** Displays the dimensions of the current feature
- **Feat Info** Provides regular feature information about the current feature in an Information window
- **Geom Check** Investigates the geometry error for the current feature
- **Fix Model** Activates Resolve mode by forcing the current feature to abort regeneration
- **Close** Closes the Model Player and enters Insert mode at the current feature
- **Finish** Closes the Model Player and returns to the last feature in the model

271

**Figure 6.28(a)** Regenerate First Feature

**Figure 6.28(b)** Regenerate the Coordinate System

**Figure 6.28(c)** Feature #5

Click: [Show Dims] [Fig. 6.28(d)] > **Close** > **Ctrl+D**

**Figure 6.28(d)** Feature #5, Revolve Dimensions

## Printing and Plotting

From the File menu, you can print with the following options: scaling, clipping, displaying the plot on the screen, or sending the plot directly to the printer. Shaded images can also be printed from this menu. You can create plot files of the current object (sketch, part, assembly, drawing, or layout) and send them to the print queue of a plotter. You can configure your printer using the MS Printer Manager Print dialog box, available from the Creo Parametric 2.0 Print dialog box. If you are printing a non-shaded image, the Printer Configuration dialog box opens.

The following applies for plotting:

- Hidden lines appear as gray for a screen plot, but as dashed lines on paper.
- When Creo Parametric 2.0 plots Creo Parametric 2.0 line fonts, it scales them to the size of a sheet. It does not scale the user-defined line fonts, which do not plot as defined.
- You can use the configuration file option *use_software_linefonts* to make sure that the plotter plots a user-defined font exactly as it appears in Creo Parametric 2.0.
- You can plot a cross section from Part or Assembly mode.
- With a Pro/PLOT license, you can write plot files in a variety of formats.

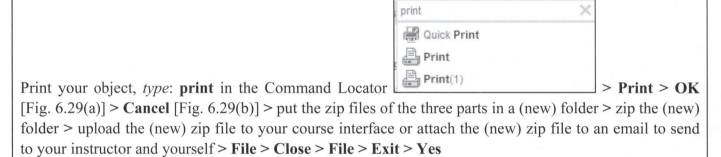

Print your object, *type*: **print** in the Command Locator > **Print** > **OK** [Fig. 6.29(a)] > **Cancel** [Fig. 6.29(b)] > put the zip files of the three parts in a (new) folder > zip the (new) folder > upload the (new) zip file to your course interface or attach the (new) zip file to an email to send to your instructor and yourself > **File** > **Close** > **File** > **Exit** > **Yes**

**Figure 6.29(a)** Print Dialog Box

**Figure 6.29(b)** MS Printer Manager Print Dialog Box

A set of projects and lectures are available at *www.cad-resources.com.*

# Lesson 7 Feature Operations

**Figure 7.1** Swing Clamp Assembly

**Figures 7.2(a-b)** Clamp Arm (Casting- Workpiece) and Clamp Arm (Machined- Design Part)

## OBJECTIVES

* Use **Copy**, **Paste**, and **Paste Special**
* Create **Ribs**
* Understand **Parameters** and **Relations**
* **Measure** geometry
* Solve **Failures**
* Create versions of the part using a **Family Table**

## REFERENCES AND RESOURCES

For **Resources** go to **www.cad-resources.com** > click on the Creo Parametric 2.0 Book cover

* Lesson 7 Lecture
* Lesson 7 3D models embedded in a PDF
* Book Projects PDF
* Project Lectures
* Creo Parametric 2.0 Quick Reference Card
* Creo Parametric 2.0 Configuration Options

## FEATURE OPERATIONS

The Clamp Arm is used in the Swing Clamp Assembly (Fig. 7.1). This lesson will cover a wide range of Creo Parametric 2.0 capabilities including: **Copy** and **Paste Special**, **Relations**, **Parameters**, **Failures**, **Family Tables**, and the **Rib Tool**. You will create two versions of the Clamp Arm; one with all cast surfaces [Fig. 7.2(a)] and the other with machined ends [Fig. 7.2(b)] (using a Family Table).

# Ribs

A profile rib is a special type of protrusion designed to create a thin fin or web that is attached to a part. You always sketch a rib from a side view, and it grows about the sketching plane symmetrically or to either side. Because of the way ribs are attached to the parent geometry, they are always sketched as open sections. A trajectory rib is also available.

When sketching an open section, Creo Parametric 2.0 may be uncertain about the side to which to add the rib. Creo Parametric 2.0 adds all material in the direction of the arrow. If the incorrect choice is made, toggle the arrow direction by picking on the direction arrow on the screen.

A **profile ribs** must "see" material everywhere it attaches to the part; otherwise, it becomes an unattached feature. There are two types of profile ribs: straight [Fig. 7.3(a)] and rotational. The type is automatically set according to the attaching geometry (planar or curved).

**Trajectory ribs** [Fig. 7.3(b)] are most often used to strengthen plastic parts that include a base and a shell or other hollow area between pocket surfaces. The pocket surface and base must consist of solid geometry. Create a trajectory rib by sketching the rib path between pocket surfaces, or by selecting an existing sketch. The rib has a top and a bottom. The bottom is the end that intersects the part surface. The sketch plane that you select defines the top surface of the rib. The rib geometry's side surfaces extend to the next surface encountered. The rib sketch can contain open, closed, self-intersecting, or multiple loops.

A Trajectory Rib feature is one trajectory, which can include any number of segments in any shape. The feature can also include a round for each edge and a draft. You can define the draft, rounds, and rib width in the graphics window or on the Trajectory Rib tab. You can separate rounds as a standalone feature that can be redefined as another feature.

**Figure 7.3(a)** Straight and Rotational Ribs    **Figure 7.3(b)** Trajectory Rib

# Relations

Relations (also known as parametric relations) are user-defined equations written between symbolic dimensions and parameters. Relations capture design relationships within features or parts, or among assembly components, thereby allowing users to control the effects of modifications on models.

Relations can be used to control the effects of modifications on models, to define values for dimensions in parts and assemblies, and to act as constraints for design conditions (for example, specifying the location of a hole in relation to the edge of a part). They are used in the design process to describe conditional relationships between different features of a part or an assembly.

Relations can be used to provide a value for a dimension. However, they can also be used to notify you when a condition has been violated, such as when a dimension exceeds a certain value. There are two basic types of relations, equality and comparison.

An equality relation equates a parameter on the left side of the equation to an expression on the right side. This type of relation is used for assigning values to dimensions and parameters. The following are a few examples of equality relations:

$$d2 = 25.500 \qquad d8 = d4/2 \qquad d7 = d1+d6/2 \qquad d6 = d2*(sqrt(d7/4.0+d4))$$

A comparison relation compares an expression on the left side of the equation to an expression on the right side. This type of relation is commonly used as a constraint or as a conditional statement for logical branching. The following are examples of comparison relations:

**d1 + d2 > (d3 + 5.5)**          Used as a constraint

**IF (d1 + 5.5) > = d7**          Used in a conditional statement

## Parameter Symbols

Four types of parameter symbols are used in relations:

- **Dimensions** These are dimension symbols, such as **d8, d12**.
- **Tolerances** These are parameters associated with **±** symmetrical and plus-minus tolerance formats. These symbols appear when dimensions are switched from numeric to symbolic.
- **Number of Instances** These are integer parameters for the number of instances in a direction of a pattern.
- **User Parameter** These can be parameters defined by adding a parameter or a relation (e.g., **Volume = d3 * d4 * d5**).

## Operators and Functions

The following operators and functions can be used in equations and conditional statements:

## Arithmetic Operators

| | |
|---|---|
| + | **Addition** |
| − | **Subtraction** |
| / | **Division** |
| * | **Multiplication** |
| ^ | **Exponentiation** |
| ( ) | **Parentheses for grouping** [for example, **(d0 = (d1–d2)*d3)**] |

## Assignment Operators

| | |
|---|---|
| = | **Equal to** |

The = (equals) sign is an assignment operator that equates the two sides of an equation or relation. When it is used, the equation can have only a single parameter on the left side.

## Comparison Operators

Comparison operators are used whenever a TRUE/FALSE value can be returned. For example, the relation **d1 >= 3.5** returns TRUE whenever d1 is greater than or equal to **3.5**. It returns FALSE whenever **d1** is less than **3.5**. The following comparison operators are supported:

| | |
|---|---|
| == | **Equal to** |
| > | **Greater than** |
| >= | **Greater than or equal to** |
| !=, <>,~= | **Not equal to** |
| < | **Less than** |
| <= | **Less than or equal to** |

| | Or |
|---|---|
| & | And |
| ~, ! | Not |

## Mathematical Functions

The following operators can be used in relations, both in equations and in conditional statements. Relations may include the following mathematical functions:

| cos ( ) | cosine |
|---|---|
| tan ( ) | tangent |
| sin ( ) | sine |
| sqrt ( ) | square root |
| asin ( ) | arc sine |
| acos ( ) | arc cosine |
| atan ( ) | arc tangent |
| sinh ( ) | hyperbolic sine |
| cosh ( ) | hyperbolic cosine |
| tanh ( ) | hyperbolic tangent |

# Failures

Sometimes model geometry cannot be constructed because features that have been modified or created conflict with or invalidate other features. This can happen when the following occurs:

- A protrusion is created that is unattached and has a one-sided edge.
- New features are created that are unattached and have one-sided edges.
- A feature is resumed that now conflicts with another feature (i.e. two chamfers on the same edge).
- The intersection of features is no longer valid because dimensional changes have moved the intersecting surfaces.
- A relation constraint has been violated.

## Resolve Feature

After a feature fails, Creo Parametric 2.0 enters Resolve Feature mode. Use the commands in the RESOLVE FEAT menu to fix the failed feature:

- **Undo Changes** Undo the changes that caused the failed regeneration attempt, and return to the last successfully regenerated model.
- **Investigate** Investigate the cause of the regeneration failure using the Investigate submenu.
- **Fix Model** Roll the model back to the state before failure and select commands to fix the problem.
- **Quick Fix** Choose an option from the **QUICK FIX** menu, the options are as follows:

  o **Redefine** Redefine the failed feature.
  o **Reroute** Reroute the failed feature.
  o **Suppress** Suppress the failed feature and its children.
  o **Clip Supp** Suppress the failed feature and all the features after it.
  o **Delete** Delete the failed feature.

# Failed Features

If a feature fails during creation and it does not use the dialog box interface, Creo Parametric 2.0 displays the FEAT FAILED menu with the following options:

- **Redefine** Redefine the feature
- **Show Ref** Display the SHOW REF menu so you can see the references of the failed feature. Creo Parametric 2.0 displays the reference number in the Message Window.
- **Geom Check** Check for problems with overlapping geometry, misalignment, and so on
- **Feat Info** Get information about the feature

If a feature fails, you can redisplay the part with all failed geometry highlighted in different colors. Creo Parametric 2.0 displays the corresponding error messages in an Information Window. Features can fail during creation for the following reasons:

- **Overlapping geometry** A surface intersects itself. If Creo Parametric 2.0 finds a self-intersecting surface, it does not perform any further surface checks.
- **Surface has edges that coincide** The surface has no area. Creo Parametric 2.0 highlights the surface in red.
- **Inverted geometry** Creo Parametric 2.0 highlights the inverted geometry in purple and displays an error message.
- **Bad edges** Creo Parametric 2.0 highlights bad edges in blue and displays an error message.

# Family Tables

**Family Tables** are effective for two main reasons: they provide a beneficial tool, and they are easy to use. You need to understand the functionality of Family Tables, and you must understand when a Family Table is required and what circumstances should promote its use.

To determine whether a model is a candidate for a Family Table: establish whether the original and the variation would ever have to co-exist at the same time (both in the same assembly, both shown in the same drawing, both with an independent Bill of Materials) and whether they should be tied together (most of the same dimensions, features, and parameters). If so, the component is a candidate for the creation of a Family Table [Figs. 7.4(a-b)], otherwise, the model may be a candidate for copying to an independent model.

**Figure 7.4(a)** Screw Family

**Figure 7.4(b)** Screws

**Figure 7.5** Clamp Arm Workpiece Drawing

## Clamp Arm

The Clamp_Arm is modeled in two stages. Model the casting (Fig. 7.5) using the casting detail, and then use the machine detail to complete the last (machined) features. The last step will be to create a Family Table with an instance that suppresses the machined features. By having a *casting part* (***workpiece***) and a separate but almost identical *machined part* (***design part***) you can create an operation for machining and an NC sequence. During the manufacturing process, *you merge the workpiece into the design part* and create a ***manufacturing model***.

The difference between the two files is the difference between the volume of the workpiece/casting part and the volume of the design part/machined part. The removed volume can be seen as *material removal* when you are performing an **NC Check** operation on the manufacturing model. If the machining process gouges the part, the gouge will display the interference. The cutter location can also be displayed as an animated machining process.

Launch **Creo Parametric 2.0** > **File** > **Manage Session** > **Select Working Directory** > select your working directory > **OK** > **File** > **New** > *type* **clamp_arm** > **OK** > **File** > **Options** > **Configuration Editor** > **Import/Export** > **Import configuration file** > **clamp.pro** *(open your previously created option file from Lesson 6)* > **Open** > **OK** > **No** > **File** > **Prepare** > **Model Properties** > Material **change** > **steel.mtl** > > **OK** > **Close** > change the coordinate system name to **clamp_arm_csys** > **Enter** > in the Graphics Window, **LMB** to deselect > > **OK** > **View** tab > *on* > *on* > *off* > **Model** tab > pick on datum **FRONT** to pre-select it > [Fig. 7.6(a)] > in the Graphics Window, press **RMB** > **Circle** > sketch a circle > move the mouse off of the circle > **MMB** to end the current tool > double-click on the diameter dimension > *type* **2.00** > **Enter** [Fig. 7.6(b)] > > click in the depth value field > *type* **1.75** > **Enter** [Fig. 7.6(c)] > > **Ctrl+D** > **Ctrl+S** > **OK**

**Figure 7.6(a)** Sketch Plane

**Figure 7.6(b)** Completed Circle

**Figure 7.6(c)** Extrusion Preview

281

With the extrusion *selected/highlighted*, click: [Copy] > [Delete] > [Paste Special] [Fig. 7.7(a)] > [Dependent copy] > [✔ Apply Move/Rotate transformations to copies] > **OK** > [Transformations] tab > pick datum **RIGHT** as the Direction reference > **LMB** on the drag handle and move to the right > when the value is **5.000**, release the **LMB** *(OR- in the value field: type 5 > Enter)* [Fig. 7.7(b)] > [✔] > [💾] > **OK** > [🔍]

**Figure 7.7(a)** Paste Special

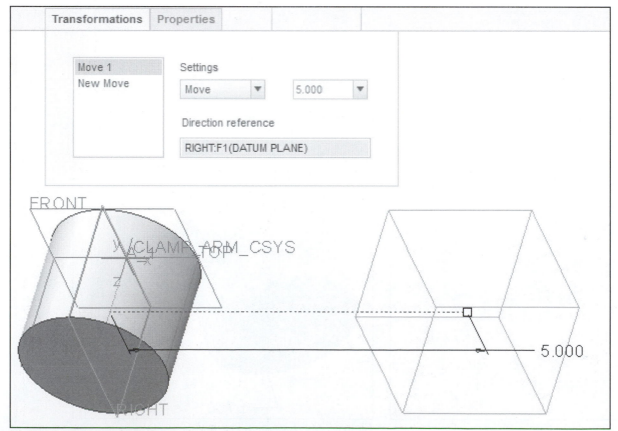

**Figure 7.7(b)** Transform the Copied Feature

282

Double-click on the copied feature > double-click on the height dimension > *type* **.875** > **Enter** > double-click on the diameter dimension > *type* **1.50** [Fig. 7.7(c)] > **Enter** > press **RMB** > **Exit Edit Dimension** > in the Graphics Window, **LMB** to deselect [Fig. 7.7(d)] > **Ctrl+S** > **MMB**

**Figure 7.7(c)** Modify the Dimensions *(double-click on each dimension and change to the design size)*

**Figure 7.7(d)** Regenerated Model

Pick on the **FRONT** datum plane > ⧠ Plane > Translation 0.4375 ▼ > **Enter** > (Fig. 7.8) > **OK**

**Figure 7.8** Offset Datum Plane

283

With datum **DTM1** *selected*, click: [Extrude] > [icon] > [icon] **References** from the Ribbon Setup Group [Figs. 7.9(a-b)] > select one arc from both circles [Fig. 7.9(c)] > **Ctrl+D** [Fig. 7.9(d)] > **Close** > [icon]

**Figure 7.9(a)** References

**Figure 7.9(b)** References Dialog Box

**Figure 7.9(c)** Add Reference

**Figure 7.9(d)** Four References

Press: **RMB** > **Construction Centerline** > create a horizontal centerline [Fig. 7.9(e)] > **MMB** > **LMB** > press **RMB** > **Circle** > pick on the center of a circular reference and then on its circular reference [Fig. 7.9(f)] > repeat on the opposite end [Fig. 7.9(g)] > **MMB** to end the Circle Tool > **Ctrl+D** [Fig. 7.9(h)]

**Figure 7.9(e)** Add a Centerline

**Figure 7.9(f)** Create a Circle

**Figure 7.9(g)** Create the Second Circle *(No weak dimensions will appear)*

**Figure 7.9(h)** Completed Circles

Click: [icon] > [▼] next to [icon] > [Line Tangent] create a line tangent between the circular extrusions [Fig. 7.9(i)] > create the bottom tangent line [Fig. 7.9(j)] > **MMB** > **LMB** to deselect > **Ctrl+D** [Fig. 7.9(k)] > **View** tab > [icons] *off* > **Sketch** tab > [icon] **Sketch View**

**Figure 7.9(i)** First Tangent Line

**Figure 7.9(j)** Second Tangent Line

**Figure 7.9(k)** 3D Sketch

Click: [Delete Segment] > press **LMB** and draw a spline through the unwanted entities on the right side [Fig. 7.9(l)] > release the **LMB** > draw a spline through the unwanted entities on the left side [Fig. 7.9(m)] > draw a spline through the tiny leftover piece on the lower-left side between the tangent position and the vertical reference [Fig. 7.9(n)] > repeat for the upper-left side [Fig. 7.9(o)]

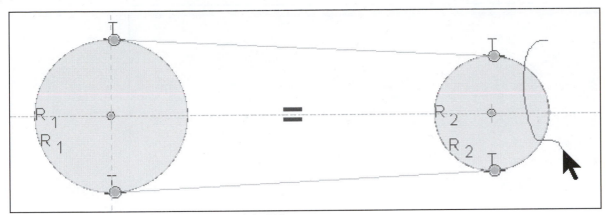

**Figure 7.9(l)** Trim Unwanted Entities on Right Side

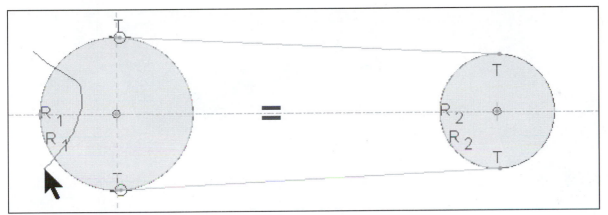

**Figure 7.9(m)** Trim Unwanted Entities on Left Side

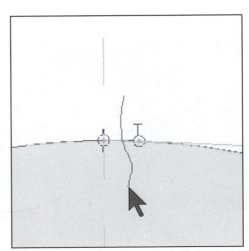

**Figure 7.9(n)** Trim Leftover on Bottom          **Figure 7.9(o)** Trim Leftover on Top

Press: **MMB** (section will shade if done correctly) [Fig. 7.9(p)] > **Ctrl+D** [Fig. 7.9(q)] > ✓ [Fig. 7.9(r)] > place the pointer over the drag handle > press **RMB** > **Symmetric** (note ⊟ in the Dashboard) > modify depth value to **.375** [ ▢ ⌒ ⊟ ▾ | 0.375 | ▾ ⅍ ⬜ ⊏ ] > **Enter** [Fig. 7.9(s)]

**Figure 7.9(p)** Completed Sketch [Note that there are no dimensions and four **T**'s (tangent constraints are displayed)]

**Figure 7.9(q)** 3D Sketch

**Figure 7.9(r)** Preview

**Figure 7.9(s)** Preview Symmetric **.375** Web Feature

Click: [✓] [Fig. 7.9(t)] > **View** tab > [icons] *on* > [icons] *on* > **Ctrl+S** > **Enter** > in the Model Tree, click: [▶] next to each feature to expand > select the sketch **Section 1** *(your name may be different)* for Extrude 3 [Fig. 7.9(u)] > in the Graphics Window, **LMB** to deselect

**Figure 7.9(t)** Completed Web Feature

**Figure 7.9(u)** Model Tree

Click: **Model** tab > select datum **TOP** > ▼ next to 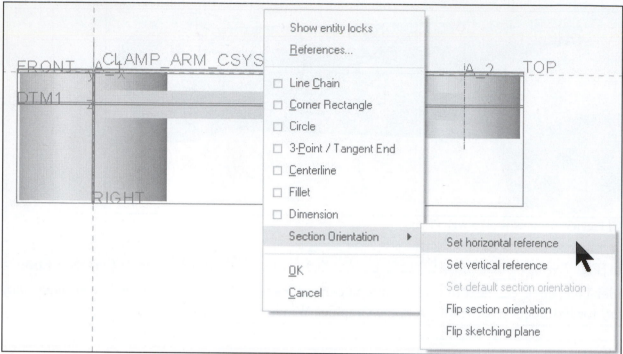 **Rib** ▼ > 🔺 Profile Rib > 🔲 > in the Graphics Window, press **RMB** > **Section Orientation** > **Set horizontal reference** [Fig. 7.10(a)] > pick on the **Line (Reference)** [Figs. 7.10(b-c)] > press **RMB** > **References** [Fig. 7.10(d)]

**Figure 7.10(a)** Profile Rib Tool and its Section Orientation

**Figure 7.10(b)** Flip the Viewing Direction

**Figure 7.10(c)** New Orientation

**Figure 7.10(d)** References Dialog Box

Click:  > add three new references, pick on the inside vertical edge of both cylindrical extrusions and the top edge of the web extrusion [Figs. 7.10(e-f)] > **Solve** > **Close** > press **RMB** > **Line Chain** draw one angled line from one vertical reference to the other [Fig. 7.10(g)] (zoom in if necessary to pick the vertical surface/edge, not the endpoint) > move the pointer > **MMB** > **MMB** [Fig. 7.10(h)] > **LMB** to deselect

**Figure 7.10(e)** New References

**Figure 7.10(f)** References Dialog Box

**Figure 7.10(g)** Sketch the Line

Press: **RMB > Dimension** add vertical dimensions from the edge reference to the end points of the angled line [Fig. 7.10(i)] > move the pointer > **MMB > MMB** > double-click on each dimension and modify the values (**.9875** and **.15**) [Fig. 7.10(j)] > **Ctrl+D** > pick on the part name in the Model Tree [Fig. 7.10(k)]

**Figure 7.10(h)** Dimensions as Sketched *(your values will be different)*

**Figure 7.10(i)** Create New dimensions *(your values will be different)*

**Figure 7.10(j)** Modify the Dimensions to the Design Values

**Figure 7.10(k)** Rib Sketch in Standard Orientation

Click:  > pick on the arrow to flip the direction of the rib creation towards the part [Fig. 7.10(l)] > press **LMB** on a drag handle > move the pointer until a rib thickness of **.375** is displayed [Fig. 7.10(m)] > release the **LMB** [image] *(or type in the value and press Enter)* > [✓] > **Ctrl+S** > **OK** [Fig. 7.10(n)] > **LMB** in the Graphics Window to deselect

*(Note: clicking [image] in the **Dashboard** of the **Profile Rib Tool** toggles the rib thickness from centered (both sides), to the right side, or to the left side of the sketch plane.)*

**Figure 7.10(l)** Flip Arrow (with the mouse pointer on the arrow, arrow color changes)

**Figure 7.10(m)** Previewed Rib

**Figure 7.10(n)** Profile Rib

293

Click:  from the Navigator > **Tree Filters** > Display: toggle all *on* > Feature Types: [Round ▾] [ ] ... 

[✓] Auto Round Member > **OK** > [▾] next to [Round ▾] > [Round / Auto Round] > [0.040 ▾] > **Enter** > **Scope** tab > **Solid Geometry** > [✓] [Fig. 7.11(a)] > **Auto-Round Player** plays [Figs. 7.11(b-c)]

**Figure 7.11(a)** Scope Tab

**Figure 7.11(b)** Auto-Round Player

**Figure 7.11(c)** Auto Round

The "machined" features of the part can be created with cuts to "face" the ends of the cylindrical protrusions. A standard tapped hole, and a through hole with two countersinks are also added as "machined" features. You will be using the dimensions from the machine drawing (Fig. 7.12).

**Figure 7.12** Clamp Arm Machining Drawing

Click: **LMB** to deselect > **Ctrl+S** > **OK** > [ Extrude ] > in the Graphics Window, press **RMB** > **Remove Material** > press **RMB** > **Define Internal Sketch** > pick datum **TOP** > **Flip** [Fig. 7.13(a)] > **Sketch** > [ ] **Sketch View** > **View** tab > [icons] *off* > [icons] *off* > **Display Style** > **Shading**

**Figure 7.13(a)** Flip Viewing Direction

295

Click: **Sketch** tab > in the Graphics Window, press **RMB** > **References** > pick the outside vertical edge of both cylindrical extrusions [Fig. 7.13(b)] > **Solve** > **Close** > press **RMB** > **Line Chain** > zoom in as needed > draw a horizontal line from one vertical reference to the other > move the pointer > **MMB** > **MMB** > **LMB** > double-click on the dimension > *type* **.100** [Fig. 7.13(c)] > **Enter** > move the pointer off of the dimension > **LMB** to deselect > press **RMB** > **OK** > **Ctrl+D** > press **RMB** > **Flip Material Side** [Fig. 7.13(d)] > **Options** tab > Side 1: **Blind** > **Through All** > Side 2: **None** > **Through All**

**Figure 7.13(b)** Add References

**Figure 7.13(c)** Draw a Line and Modify the Dimension

**Figure 7.13(d)** Material Removal Direction

Click:  > **View** tab > ⊡ on > ⊡ on [Figs. 7.13(e-f)] > in the Graphics Window, **LMB** to deselect

**Figure 7.13(e)** Display     **Figure 7.13(f)** Datum Planes and Tags Displayed

Click: **Model** tab > **MMB** rotate the model to see its bottom surface cut > pick on the cut > move the cursor a small distance and pick again (the cut *surface* will highlight) [Fig. 7.14(a)] > `⊡ Plane` > **Offset** > **Offset** > **Through** [Fig. 7.14(b)] > **OK** [Fig. 7.14(c)] > **LMB** to deselect > **Ctrl+D** > **Ctrl+S** > **OK**

**Figure 7.14(a)** Select the Surface     **Figure 7.14(b)** Datum Plane Through

**Figure 7.14(c)** Completed Datum Plane

Click: **View** tab >  *on* > pick **DTM2** from the Model Tree > **RMB** > **Properties** > Name- *type* **A** > > **On Datum** > **OK** > pick **TOP** > **RMB** > **Properties** > Name- *type* **B** > > **On Datum** > **OK** > pick **RIGHT** > **RMB** > **Properties** > Name- *type* **C** > > **On Datum** > **OK** > press the **Ctrl** key > pick **FRONT** and **DTM1** > release the **Ctrl** key > **RMB** > **Hide** [Fig. 7.15] > **Model** tab > pick datum **A** > Plane > **LMB** on the drag handle and move towards the top of the rib > release the **LMB** > Offset Translation **1.550** [Fig. 7.16(a)] > **Enter** > **OK** [Fig. 7.16(b)] > **LMB**

**Figure 7.15** Set Datum Tag Annotations

**Figure 7.16(a)** Offset Datum

**Figure 7.16(b)** Completed Offset Datum *(your id values may be different)*

Click:  **Extrude** > > **Placement** tab > **Define** > pick **DTM2** > Reference- pick datum **C** > Orientation **Right** [Fig. 7.17(a)] > **Sketch** to close the Sketch dialog box > pick on the circular surface of the large protrusion to add as a reference [Figs. 7.17(b-c)] > pick **F1(C)** in the References dialog box > **Delete** the extra reference > **Solve** > **Close** > > **No Hidden** [Fig. 7.17(d)]

**Figure 7.17(a)** Sketch Dialog Box

**Figure 7.17(b)** References

**Figure 7.17(c)** Select the Cylindrical Surface

**Figure 7.17(d)** Circular Reference

Press: **RMB** > **Circle** > sketch a circle by picking the center [Fig. 7.17(e)] and one edge (on surface reference) [Fig. 7.17(f)] *(the circle is locked into the reference, no dimensions will be displayed; if you get dimensions, undo and redo the circle)* > **MMB** [Fig. 7.17(g)] >  [Fig. 7.17(h)] > > > > **Shading** [Fig. 7.17(i)] > > **LMB** to deselect > > **Shading With Edges** [Fig. 7.17(j)]

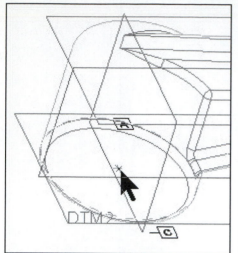

**Figure 7.17(e)** Pick the Center

**Figure 7.17(f)** Pick the Edge Reference

**Figure 7.17(g)** Completed Circle

**Figure 7.17(h)** Cut

**Figure 7.17(i)** Previewed Cut

**Figure 7.17(j)** Completed Cut

Click: **DTM2** in the Model Tree > **RMB** > **Hide** > [Extrude] > [icon] > [icon] > [icon] > **Placement** tab >

**Define** > [Datum icon] > [Plane icon] **Plane** > pick datum **A** > **LMB** on the drag handle and move > double-click on the dimension > *type* **.725** > **Enter** [Fig. 7.17(k)] > **OK** > Reference **B:F2(DATUM PLANE)** > Orientation **Top** [Fig. 7.17(l)] > **Sketch** > press **RMB** > **References** > hold down the **Ctrl** key and select **F1** and **F2** > **Delete** > [cursor icon] > pick on the circular surface of the small protrusion [Figs. 7.17(m-n)] > **Solve** > **Close**

**Figure 7.17(k)** Creating an Internal Datum Plane Offset from Datum A for Sketching

**Figure 7.17(l)** Sketch Plane and Orientation

**Figure 7.17(m)** Select the Surface

**Figure 7.17(n)** Reference

Click:  > sketch a circle by picking the center [Fig. 7.17(o)] and the surface reference [Fig. 7.17(p)] *(the circle is locked into references, no dimensions will be displayed; if you get dimensions, undo and redo the circle)* > **MMB** [Fig. 7.17(q)] >  [Fig. 7.17(r)] >  [Fig. 7.17(s)] > in the Graphics Window, **LMB** to deselect > **Ctrl+S** > **OK**

**Figure 7.17(o)** Pick Center

**Figure 7.17(p)** Pick Reference

**Figure 7.17(q)** Shaded Section

**Figure 7.17(r)** Previewed Through All Cut

**Figure 7.17(s)** Completed Cut

302

# Measuring Geometry

Using analysis measure, you can measure model geometry with one of the following commands:

- **Length**  Displays the length of the curve or edge
- **Distance**  Displays the distance between two entities
- **Angle**  Displays the angle between two entities
- **Diameter**  Displays the diameter of the surface
- **Radius** Displays the radius of the surface
- **Area**  Displays the area of the selected surface, quilt, facets, or an entire model
- **Volume**  Displays the volume of the object
- **Transform** Displays the transformation matrix between two coordinate systems.

Click: **Analysis** tab > **Measure** > **Distance** > in the Measure: dialog box; click: [▼] to expand or [▲] to collapse the dialog box regions > pick the top surface of the large circular protrusion [Fig. 7.18(a)] > press and hold **Ctrl** > place the cursor over the bottom and click **RMB** until it highlights [Fig. 7.18(b)] > **LMB** to select > release the **Ctrl** key > select a **Reference** > **RMB** > **Remove All** > repeat the process to measure the shorter end [Fig. 7.18(c)] (the design sizes are **1.550** and **.725** respectively) > [X]

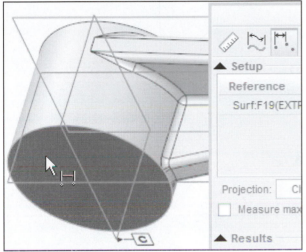

**Figure 7.18(a)** Select Top Surface

**Figure 7.18(b)** Select Bottom Surface- Distance **1.55000**

**Figure 7.18(c)** Distance **.725000**

303

## Editing the Model

During the design of a component there are modifications made to the design. The ability to make changes without causing failures is important. "Flexing" the model; changing and editing dimension values to see if the model integrity withstands these modifications, establishes your designs robustness.

Click: **Auto Round 1** feature in the Model Tree > **RMB** > **Edit** > double-click on the dimension value > *type* **.08** > **Enter** > move the pointer off of the dimension > **LMB** to complete the edit > move the pointer > **LMB** to regenerate and deselect > double-click on the **Rib** > double-click on the **.988** dimension value > *type* **.90** > **Enter** > press **RMB** > **Exit Edit Dimension** > **LMB** > double-click on the **Rib** > drag a handle slightly until the value is **1.000** *(be patient- may take several minutes)* [Fig. 7.19(a)] > release the

LMB > move the pointer into the Graphics Window > **LMB** [Figs. 7.19(b-c)] >  **Undo** (may have to

| Surf:...ND_1) | ⊖ | |
|---|---|---|
| Diameter | 0.0800000 | in |
| Radius | 0.0400000 | in |

select several times) > **Measure** > **Diameter** > select the round > [table above] > **☒** > **Model** tab

**Figure 7.19(a)** Modify the **.375** Dimension by dragging a handle to **1.00**

**Figure 7.19(b)** Regenerate

**Figure 7.19(c)** Flexed Part

Click: 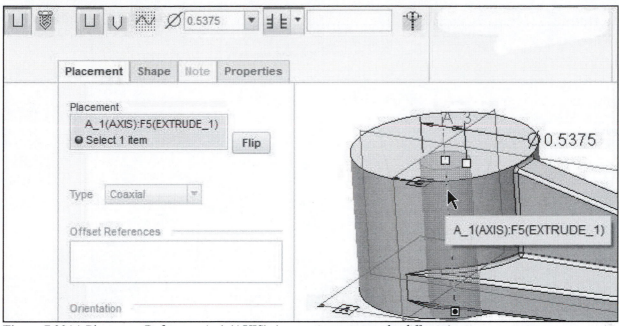 > **OK** > **View** tab > ☐ *on* > ☐ *on* > **Model** tab > `⊔ Hole` > `⊥` > `⊟` > `⌀ 0.5375000 ▼` > **Placement** tab > pick the Axis [Fig. 7.20(a)] > hold down the **Ctrl** key > pick the top surface [Fig. 7.20(b)] > release the **Ctrl** key > **Shape** tab [Fig. 7.20(c)] > **Enter** > **LMB** to deselect

**Figure 7.20(a)** Placement Reference A_1 (AXIS) *(your axis name may be different)*

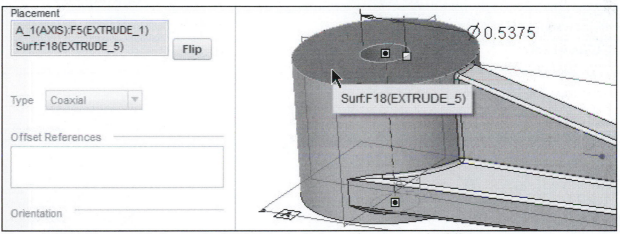

**Figure 7.20(b)** Placement Reference Surf:F18(Extrude_5) *(your id names may be different)*

**Figure 7.20(c)** Shape Tab

305

Click: > **OK** > [Chamfer] > [icons] D x D ▼ D 0.050 ▼ > **Sets** tab > hold down the **Ctrl** key > pick the top and bottom edges of the hole [Fig. 7.20(d)] > release the **Ctrl** key > [✓] > [💾] > **OK**> **LMB** to deselect

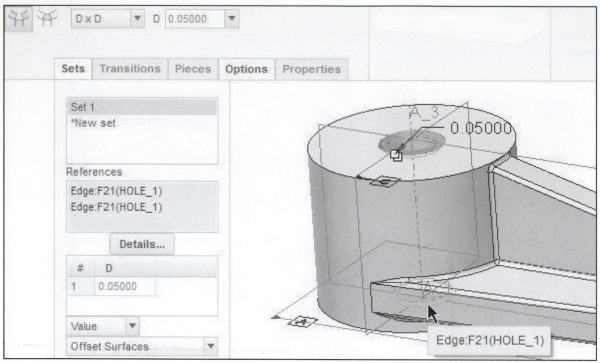

**Figure 7.20(d)** Chamfer Edges

Click: [🔩 Hole] > pick axis **A_2** [Fig. 7.21(a)]

**Figure 7.21(a)** Placement Axis *(your axis name may be different)*

Press and hold the **Ctrl** key > pick the top surface of the small cylinder > release the **Ctrl** key > **Placement** tab > [icon] **Create standard hole** > [1/2-13 ▼] > [icon ▼] > [icon] > [icon] **Adds countersink** [Fig. 7.21(b)] > **Shape** tab > [● Thru Thread] > [☑ Exit Countersink] > [0.5625 ▼] top and bottom chamfer [Fig. 7.21(c)]

**Figure 7.21(b)** Placement Surface

**Figure 7.21(c)** Shape Tab

Click: **Note** tab [Fig. 7.21(d)] >  (if you get a failure, click the Shape tab and check the options and dimensions) >  **Settings > Tree Filters >** ☑ Annotations [Fig. 7.21(e)] > **OK >** ▼ ⚙ Hole 2 / A☰ Note_0 [Fig. 7.21(f)] > **Ctrl+D > Ctrl+S > OK >** in the Graphics Window, **LMB** to deselect

**Figure 7.21(d)** Note Tab

**Figure 7.21(e)** Model Tree Items

**Figure 7.21(f)** Note Selected in Model Tree and Displayed on the Model

Slowly click twice on the **Extrude 3** feature in the Model Tree > *type* **WEB** as the new name > **Enter** > in the Graphics Window, **LMB** to deselect > Write a relation to keep the thickness of the rib the same as that for the web. With the **Ctrl** key pressed, pick on the WEB and the rib in the Model Tree > release the **Ctrl** key > **RMB** > **Edit** > place the mouse pointer over each dimension [Fig. 7.22(a)] and note the **d** values [Fig. 7.22(b)] > **Tools** tab > d= Relations > *type (or pick from the model)* **d8=d7** [Fig. 7.22(c)] *(your "d" values may be different)* > ☑ **Execute/Verify** > **OK** > **OK** > 💾 > **OK** > **LMB** to deselect > 🖉

**Figure 7.22(a)** Web (**d7**) *[use your "d" values]*

**Figure 7.22(b)** Rib (**d8**) *[use your "d" values]*

**Figure 7.22(c)** Relations Dialog Box *(your "d" values may be different)*

Press: **Ctrl+D** > pick on the **WEB** in the Model Tree > **RMB** > **Edit** > double-click on the **.375** dimension > *type* **.60** [Fig. 7.22(d)] > **Enter** [Fig. 7.22(e)] > move the pointer into the Graphics Window > **LMB** [Fig. 7.22(f)] > slightly move the pointer > **LMB** > **Cancel** [Fig. 7.22(g)] > repeat and modify the value to **.20** > **Enter** > move the pointer > **LMB** > move the pointer > **LMB** [Fig. 7.22(h)] >  **Undo**

**Figure 7.22(d)** Edit the Web Dimension to **.60**

**Figure 7.22(e)** Failed Feature

**Figure 7.22(f)** Failed Features

⚠ Some features failed to regenerate.
Choose OK to accept the result or Cancel to undo the changes.

[ OK ]  [ Cancel ]

**Figure 7.22(g)** Failure Diagnostics Dialog Box

**Figure 7.22(h)** **.20** Dimension driving the WEB and the Rib Size

310

# Family Tables

**Family Tables** are used any time a part or assembly has several unique iterations developed from the original model. The iterations must be considered as separate models, not just iterations of the original model. In this lesson, you will add instances of a family table to create a machined part and a casting of the Clamp Arm (a version without machined cuts or holes).

You will be creating a Family Table from the generic model. The (base) model is the **Generic**. Each variation is referred to as an **Instance**. When you create a Family Table, Creo Parametric 2.0 allows you to *select dimensions,* which can vary between instances. You can also *select features* to add to the Family Table. Features can vary by being suppressed or resumed in an instance. When you are finished selecting items (e.g., dimensions, features, and parameters), the Family Table is automatically generated.

When adding features to the table; enter an **N** to suppress the feature, or a **Y** to resume the feature. Each instance must have a unique name.

Family tables are spreadsheets, consisting of columns and rows. *Rows* contain instances and their corresponding values; *columns* are used for items. The column headings include the *instance name* and the names of all of the *dimensions, parameters, features, members,* and *groups* that were selected to be in the table. The Family Table dialog box is used to create and modify family tables.

Family tables include:

- The base object (generic object or *generic*) on which all members of the family are based.
- Dimensions, parameters, feature numbers, user-defined feature names, and assembly member names that are selected to be table-driven (*items*).
    - **Dimensions** are listed by name (for example, **d125**) with the associated symbol name (if any) on the line below it (for example, depth).
    - **Parameters** are listed by name (dim symbol).
    - **Features** are listed by feature number with the associated feature type (for example, [cut]) or feature name on the line below it. The generic model is the first row in the table. Only modifying the actual part, suppressing, or resuming features can change the table entries belonging to the generic; *you cannot change the generic model by editing its row entries in the family table.*
- Names of all family members (*instances*) created in the table and the corresponding values for each of the table-driven items

Click: **Tools** tab > Family Table --the Family Table: dialog box opens [Fig. 7.23(a)]

**Figure 7.23(a)** Family Table Dialog Box

Click:  **Add/delete the table columns** > ⦿ Feature from the Add Item options > select the cuts and holes from the model or the Model Tree [Fig. 7.23(b)] *(the order in which the items are listed will determine the default order in which they will appear in the table – default column order)* > **OK** [Fig. 7.23(c)] > **Insert a new instance at the selected row** > pick on the name of the new instance **CLAMP_ARM_INST** > *type* **CLAMP_ARM_DESIGN** [Fig. 7.23(d)] > **Enter** *(adds a new instance)*

**Figure 7.23(b)** Family Items Dialog Box, Adding Features

**Figure 7.23(c)** New Family Table

**Figure 7.23(d)** Add an Instance

The second instance **CLAMP_ARM_INST** should be highlighted, *type* **CLAMP_ARM_WORKPIECE** [Fig. 7.23(e)] > pick in the cell of the first feature and change to **N** (not used) [Fig. 7.23(f)] > change all cells for the CLAMP_ARM_WORKPIECE to **N** [Fig. 7.23(g)]

**Figure 7.23(e)** Add a Second Instance

**Figure 7.23(f)** Change Feature to N (not used) *(pick on the drop down arrow for the list of choices)*

**Figure 7.23(g)** N for All Machined Features

Note: If your school or company uses the parameter "Common Name", enter the appropriate data.

Click:  **Verify instances of the family** [Fig. 7.23(h)] > **VERIFY** [Fig. 7.23(i)] > **CLOSE** [Fig. 7.23(j)] > **OK** from Family Table dialog box > **Ctrl+D** > **Ctrl+S** > **OK**

**Figure 7.23(h)** Verify Instances of the Family

**Figure 7.23(i)** Verify

**Figure 7.23(j)** Verification Status

A Family Table controls whether a feature is present or not for a given design instance, not whether a feature is displayed. The Generic is the base model [Fig. 7.23(k)] and is typically not a member of an assembly.

Click: **Family Table > CLAMP_ARM_DESIGN >** [Open] [Fig. 7.23(l)] > [Close] **Close > Family Table > CLAMP_ARM_WORKPIECE >** [Open] [Fig. 7.23(m)] > [Close] **Close** (the Generic will be the only object on your screen) **> Ctrl+S > OK**

**Figure 7.23(k)** Instance: GENERIC

**Figure 7.23(l)** Instance: CLAMP_ARM_DESIGN

**Figure 7.23(m)** Instance: CLAMP_ARM_WORKPIECE

Click: **Annotate** tab > select the features in the Model Tree [Fig. 7.24(a)] > **RMB** > **Show Annotations** > [Fig. 7.24(b)] > **Apply** > **Cancel** [Fig. 7.24(c)] > pick on a dimension > press **RMB** > **Properties** > **Move** > pick a new position > **OK** > **LMB** > pick on ⟨A⟩ > press **RMB** > **Flip** (⟨A⟩) [Fig. 7.24(d)]

**Figure 7.24(a)** Create Driving Dimension AE

| Show | Type | Name |
|:---:|:---:|:---|
| ✓ | 🗠 | d0 |
| ✓ | 🗠 | d1 |
| ✓ | 🗠 | d3 |
| ✓ | 🗠 | d4 |
| ✓ | 🗠 | d6 |
| ✓ | 🗠 | d7 |
| ✓ | 🗠 | d8 |
| ✓ | 🗠 | d9 |
| ✓ | 🗠 | d10 |
| ✓ | 🗠 | d15 |
| ✓ | 🗠 | d17 |
| ✓ | 🗠 | d20 |
| ✓ | 🗠 | d21 |
| ✓ | 🗠 | d23 |
| ✓ | 🗠 | d26 |
| ✓ | 🗠 | d29 |
| ✓ | 🗠 | d30 |
| ✓ | 🗠 | d31 |
| ✓ | 🗠 | d33 |
| ✓ | 🗠 | d34 |

**Figure 7.24(b)** Select All

**Figure 7.24(c)** Detail Tree

**Figure 7.24(d)** Repositioned Dimensions

Click:  [Fig. 7.24(e)] > **Ctrl+D** > **Ctrl+S** > **Enter** > **Ok** (if needed, from Conflicts dialog box) > **Tools** tab > **Model Intent** Group > **Relations and Parameters** [Fig. 7.25] > **File** > **Manage File** > **Delete Old Versions** > **Enter** > **File** > **Save As** > **Save a Copy** > Type ▭ > **Zip File (*.zip)** > **OK** > upload the zip file to your class interface system and email to yourself and your instructor > **File** > **Close**

**Figure 7.24(e)** Annotation Orientation Plane

**PART NAME :** CLAMP_ARM

### Features Containing Relations/Parameters

| ID | Name | Type | Actions | |
|----|------|------|---------|---|
| 6610 | — | HOLE_2 | | |

### Relation Table

| Relation | Parameter | New Value |
|----------|-----------|-----------|
| | Relations for CLAMP_ARM: | |
| | **Initial Relations** | |
| d8 =d7 | D8 | 3.750000e-01 |

**Figure 7.25** Relations and Parameters

# Creo Parametric 2.0 MANUFACTURING

You now have two separate models, a casting (workpiece) and a machined part (design part). During the manufacturing process, the workpiece is merged (assembled) into the design part thereby creating a manufacturing model [Fig. 7.26(a)]. The difference between the two objects is the difference between the volume of the casting and the volume of the machined part. The manufacturing model is used to machine the part [Fig. 7.26(b)]. If you have the manufacturing module you could now virtually "machine" the part.

**Figure 7.26(a)** Manufacturing Model

**Figure 7.26(b)** Facing

A complete set of extra projects are available at *www.cad-resources.com*

**Figure 8.1(a)** Swing Clamp Assembly

## OBJECTIVES

- **Assemble** components to create an assembly
- Create a **subassembly**
- Understand and use a variety of **Assembly Constraints**
- **Modify** a component constraint
- Check for **clearance** and **interference**

## REFERENCES AND RESOURCES

For **Resources** go to **www.cad-resources.com** > click on the Creo Parametric 2.0 Book cover

- Lesson 8 Lecture
- Lesson 8 3D models embedded in a PDF
- Book Projects PDF
- Project Lectures

## ASSEMBLY CONSTRAINTS

**Assembly mode** allows you to place together components and subassemblies to create an assembly [Fig. 8.1(a)]. Assemblies [Fig. 8.1(b)] can be modified, reoriented, documented, or analyzed. An assembly can be assembled into another assembly, thereby becoming a subassembly.

**Figure 8.1(b)** Swing Clamp (CARRLANE at **www.carrlane.com**)

## Placing Components

One new addition to PTC's software is the **3D Dragger**. The 3D Dragger is a graphical tool for precision handling of assembly components. The *axes* are called **draggers**. Pull or rotate the draggers to make changes to the components position and orientation. The draggers enable movement in one or more  *degrees of freedom* (DOFs). They support linear, planar, free, translation, and angular movements. You can pull a dragger along a 2D or 3D trajectory. When you move a geometric entity using a dragger, the relative distances from the start point of the entity appear in the graphics window as you pull the dragger. You can hold down the SHIFT key while pulling a dragger to snap the dragger to a geometric reference.

To assemble components, use: **Add a component to the assembly**. After selecting a component from the Open dialog box, the dashboard (Fig. 8.2) opens and the component appears in the assembly window. Alternatively, you can select a component from a browser window and drag it into the Creo Parametric 2.0 window. If there is an assembly in the window, Creo Parametric 2.0 will begin to assemble the component into the current assembly. Using icons in the dashboard, you can specify the screen window in which the component is displayed while you position it. You can change window options at any time using: **Show component in a separate window while specifying constraints** or **Show component in the assembly window while specifying constraints**.

**Figure 8.2** Component Placement Dashboard

**Figure 8.3(a)** Swing Clamp Main Assembly

**Figure 8.3(b)** Swing Clamp Sub-Assembly

## Swing Clamp Assembly

The parts required in this lesson are from this text. *If you have not modeled these parts previously, please do so before you start the following systematic instructions.* The other components required for the assembly are standard *off-the-shelf* hardware items that you can get by accessing a company catalog. The **Flange Nut**, the **3.50 Double-ended Stud**, and the **5.00 Double-ended Stud** are standard items. The **Clamp Plate** component is the first component of the main assembly and will be modeled later when completing the main assembly [Fig. 8.3(a)] using the *top-down design* approach.

Because you will be creating the sub-assembly [Fig. 8.3(b)] using the *bottom-up design* approach, all the components must be available before any assembling starts. *Bottom-up design* means that existing parts are assembled, one by one, until the assembly is complete. The assembly starts with a set of default datum planes and a coordinate system. The parts are constrained to the datum features of the assembly. The sequence of assembly will determine the parent-child relationships between components.

*Top-down design* is the design of an assembly where one or more component parts are created in Assembly mode as the design unfolds. Some existing parts are available, such as standard components and a few modeled parts. The remaining design evolves during the assembly process. The main assembly will involve creating one part using the *top-down design* approach.

Regardless of the design method, the assembly default datum planes and coordinate system should be on their own separate *assembly layer*. Each part should also be placed on separate assembly layers; the part's datum features should already be on *part layers*.

Before starting the assembly, you will be modeling each part or retrieving *standard parts* and saving them under unique names into *your* working directory. *Unless instructed to do so, **do not use the library parts directly in the assembly**.* Start this process by retrieving the standard parts [Figs. 8.3(c-e)].

**The three standard parts are also available at *http://www.cad-resources.com > pick on the appropriate book cover > download the commercial or academic versions > drag and drop the (each) file into an empty Graphics Window***

**Figures 8.3(c-e)** Standard Parts from Carr Lane

Launch **Creo Parametric 2.0** > **Select Working Directory** > select the working directory > **OK** > open the Browser *(if needed)* > `3DModelSpace` `×` ( `http://www.3dmodelspace.com/ptc` ) > *type* www.carrlane.com in the address bar

`http://www.carrlane.com/` > **Enter** [Fig. 8.4(a)]

*Alternative method 1: download parts from http://www.cad-resources.com*
*Alternative method 2: model the three parts*

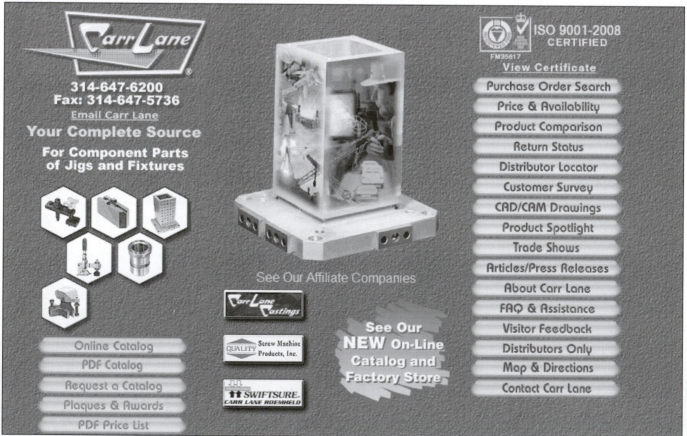

**Figure 8.4(a)** Carr Lane (The Website may have since been updated)

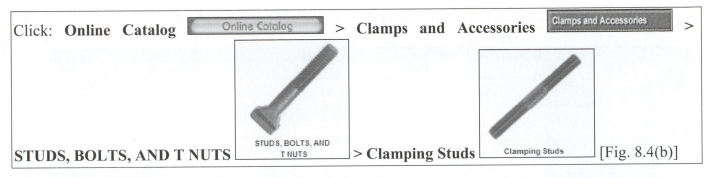

Click: **Online Catalog** [Online Catalog] > **Clamps and Accessories** [Clamps and Accessories] >

**STUDS, BOLTS, AND T NUTS** [STUDS, BOLTS, AND T NUTS] > **Clamping Studs** [Clamping Studs] [Fig. 8.4(b)]

| Features | Dimensions |

CLAMPS AND ACCESSORIES » STUDS, BOLTS, AND T NUTS » Clamping Studs

# CLAMPING STUDS

Steel & Stainless Steel

**FEATURES:** High-strength general-purpose Clamping Stud, for use with a T Nut, or for installation directly into a fixture plate. Available in steel or stainless steel. Made in USA.

**SIZES (STEEL):** Available in the following thread sizes, each in a choice of many lengths — see Dimensions page for part numbers:

| | |
|---|---|
| 1/4-20 | M6 |
| 5/16-18 | M8 |
| 3/8-16 | M10 |
| 1/2-13 | M12 |
| 5/8-11 | M16 |
| 3/4-10 | M20 |
| 7/8-9 | — |
| 1"-8 | — |

**SIZES (STAINLESS):** Available in the following thread sizes, each in a choice of many lengths — see Dimensions page for part numbers:

3/8-16
1/2-13
5/8-11

**MATERIAL (STEEL):** 1144 steel, 115,000 psi min tensile strength, 100,000 psi min yield strength, black oxide finish. DFARS and RoHS compliant.

**MATERIAL (STAINLESS):** 300-series stainless, passivated. DFARS and RoHS compliant.

For smaller stainless steel sizes, see Fixture Studs.

en_447

**Figure 8.4(b)** Carr Lane CLAMPING STUDS

323

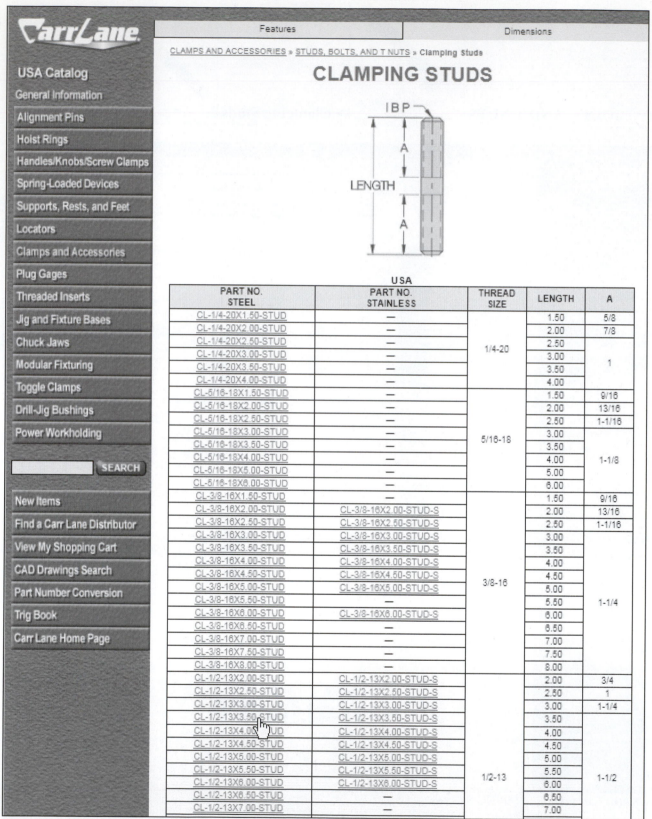

**Figure 8.4(c) CL-1/2-13X3.50-STUD**

Click: **Interactive 3D Viewer** 3D Viewer [Fig. 8.4(d)] (download the viewer (plugin) if necessary or go directly to the next command) > **CAD Downloads** CAD Downloads > 3D Download Formats: ▼ > **Pro/E** *(Creo Parametric 2.0)* **Part/Assembly (\*.prt) > Go** Go [Fig. 8.4(e)]

**Figure 8.4(d)** 3D Viewer (download the viewer if desired or allowed)

CLAMPS AND ACCESSORIES » STUDS, BOLTS, AND T NUTS » Clamping Studs

## CLAMPING STUDS

### Part No: CL-1/2-13X3.50-STUD
Price (1pc): $1.56

Enter quantity for stock check and quantity price: [        ] Go

---

### Drawing and Model Selections

Current Configuration: 12133.50stud ▼ [ Change ]

Interactive 3D Viewer    CAD Downloads

3D Download Formats:                    Version:

[ Pro/E Part/Assembly (\*.prt) ▼ ] [ 2012 ▼ ] [ Go ]

**Figure 8.4(e)** Pro/E *(Creo Parametric 2.0)* Part/Assembly (\*.prt)

325

Click: [Download Drawing] > Save [▼] > **Save as** (Save As dialog box opens) *[if this does not work, right click on* [Download Drawing] *and select "Save Target As..."]* > navigate to your Working Directory > in File name field; *type* (a new name) **Carrlane-12-13350_STUD.prt** > **Enter** > double-click on the Browser sash to close > **File** > **Open** > **Carrlane-12-13350_STUD.prt** > **Open** > **View** tab > **Appearance Gallery** > [▼] next to Clear Appearance > **Clear All Appearances** [Fig. 8.4(f)] > **Yes** > change the coordinate system name to **350_STUD** > **Enter** [Fig. 8.4(g)] > in the Graphics Window, **LMB** to deselect > **Ctrl+D** > **Ctrl+S** > **Enter** > [icon] **Close** *(you may also model the stud [Fig. 8.4(h)])*

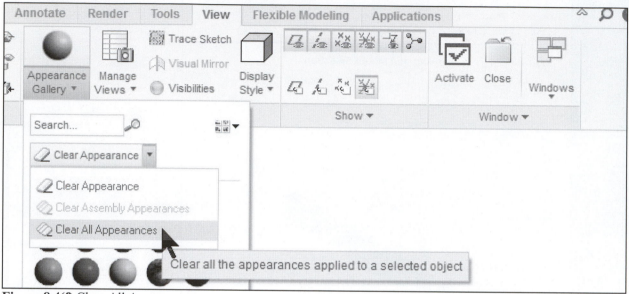

**Figure 8.4(f)** Clear All Appearances

**Figure 8.4(g)** New Coordinate System Name: **350_STUD**

**Figure 8.4(h)** Model your own stud (without threads) if the catalog is not available

Click:  open the Browser > ⬅ > ⬅ until you get to the CLAMPING STUDS Dimensions page > select CL-1/2-13X5.00-STUD > **CAD Downloads** > 3D Download Formats: **Pro/E** *(Creo Parametric 2.0)* **Part/Assembly (\*.prt)** > **Go** > **Download Drawing** > **Save** ▼ > **Save as** > navigate to your Working Directory > File name **Carrlane-12-13500_STUD.prt** > **Save** > ✕ to close Browser > 📂 > **Views** tab > **Thumbnails** [Fig. 8.4(i)] > **Carrlane-12-13500_STUD.prt** > **Open** > **View** tab > **Appearance Gallery** > ▼ next to Clear Appearance > **Clear All Appearances** > **Yes** > rename the coordinate system to **500_STUD** [Fig. 8.4(j)] > **Ctrl+D** > 💾 > **OK** > 📂 *(you may also model the stud [Fig. 8.4(k)])*

**Figure 8.4(i)** File > Open > Views Tab > Thumbnails > Carrlane-12-13500_STUD.prt

**Figure 8.4(j)** New Coordinate System Name: **500_STUD**

**Figure 8.4(k)** Double-ended Stud ∅**.500** by **5.00** Length

Click:  open Browser > ⬅ > ⬅ > ⬅ > ⬅ > ⬅ > ⬅ until CLAMPS AND ACCESSORIES

page [Fig. 8.4(l)] > **NUTS** > **Flange Nuts** [Fig. 8.4(m)]

**Figure 8.4(l)** CLAMPS AND ACCESSORIES

## FLANGE NUTS

CLAMPS AND ACCESSORIES » NUTS » Flange Nuts

FEATURES: Heavy-hex nut with a flanged face. This large bearing surface prevents damage to the part underneath. Ideal for clamping over a slot. Fully machined from solid barstock for extra strength, unlike inferior substitutes. Available in steel or stainless steel. Made in USA.

SIZES (STEEL): Available in the following thread sizes — see Dimensions page for part numbers:

| | |
|---|---|
| 1/4-20 | M6 |
| 1/4-28 | — |
| 5/16-18 | M8 |
| 5/16-24 | — |
| 3/8-16 | M10 |
| 3/8-24 | — |
| 1/2-13 | M12 |
| 1/2-20 | — |
| 5/8-11 | M16 |
| 5/8-18 | — |
| 3/4-10 | M20 |
| 3/4-16 | — |
| 7/8-9 | — |
| 1"-8 | M24 |
| 1"-14 | — |
| 1-1/4"-7 | — |

SIZES (STAINLESS): Available in the following thread sizes — see Dimensions page for part numbers:

Flange Nut

**Figure 8.4(m)** FLANGE NUTS

Click: **Dimensions** [Fig. 8.4(n)] > select `CL-3-FN` > **Interactive 3D Viewer** > **CAD Downloads** > 3D Download Formats: **Pro/E** *(Creo Parametric 2.0)* **Part/Assembly** > **Go** > **Download Drawing** > Save ⬇ > **Save as** > navigate to your Working Directory > File name, *type* **carrlane_500_fn.prt** > **Save** > double-click on the Browser sash to close > 📂 **Open** > **carrlane_500_fn.prt** > **Open** > **View** tab > **Appearance Gallery** > ⬇ next to Clear Appearance > **Clear All Appearances** > **Yes** > rename the coordinate system to **CARRLANE_500_FN** [Fig. 8.4(o)] > **Enter** > in the Graphics Window, **LMB** to deselect > **Ctrl+D** > **Ctrl+S** > **Enter** > **File** > **Close** *(you may also model the Flange Nut [Fig. 8.4(p)])*

# FLANGE NUTS

| STEEL PART NO. | STAINLESS STEEL PART NO. | A | B | C DIA | D | E |
|---|---|---|---|---|---|---|
| | CL-123 | #10-24 | 5/16 | 1/2 | 3/8 | .08 |
| | CL-125 | 1/4-20 | 5/16 | 9/16 | 7/16 | .08 |
| CL-1-FN | | | | 5/8 | 1/2 | .12 |
| | CL-126 | 5/16-18 | 3/8 | 11/16 | 9/16 | .09 |
| CL-9-FN | | | | 3/4 | | .12 |
| CL-2-FN | CL-127 | 3/8-16 | 1/2 | 7/8 | 11/16 | .12 |
| CL-3-FN | CL-128 | 1/2-13 | 11/16 | 1-1/8 | 7/8 | .20 |
| CL-4-FN | CL-129 | 5/8-11 | 13/16 | 1-3/8 | 1-1/16 | |

**Figure 8.4(n) CL-3-FN**

**Figure 8.4(o)** Flange Coordinate System Name

**Figure 8.4(p)** Flange Nut Dimensions for Modeling

Open the models for the Clamp_Arm_Design [Fig. 8.5(a)], the Clamp_Swivel [Fig. 8.5(b)], the Clamp_Ball [Fig. 8.5(c)] and the Clamp_Foot [Fig. 8.5(d)]. > Review the components and standard parts for unique **colors**, **layering**, **coordinate system naming**, and **set datum planes**. > Close all windows. Components remain "in session".

**Figure 8.5(a)** Clamp_Arm_Design

**Figure 8.5(b)** Clamp_Swivel

**Figure 8.5(c)** Clamp_Ball

**Figure 8.5(d)** Clamp_Foot

You now have eight components (two identical Clamp_Ball components are used) required for the assembly. The Clamp_Plate (Fig. 8.6) will be created using *top-down design* procedures when you start the main assembly. *All parts must be in the same working directory used for the assembly.*

**Figure 8.6** Clamp_Plate

A subassembly will be created first. The main assembly is created second. The subassembly will be added to the main assembly to complete the project. Note: the assembly and the components can have different units. Therefore, you must check and correctly set the assembly units before creating or assembling components or sub-assemblies.

Click: **Create a new model** > Assembly > Sub-type Design > Name **clamp_subassembly** > Use default template [Fig. 8.7(a)] > **OK** > Template **inlbs_asm_design** > Parameters MODELED_BY *type **your name*** > DESCRIPTION *type* **Swing Clamp sub-assembly** [Fig. 8.7(b)] > **OK** > **File** > **Prepare** > **Model Properties** > Units change > Inch lbm Second (Creo Parametric Default) > **Close** > **Close**

**Figure 8.7(a)** New Dialog Box

**Figure 8.7(b)** New File Options Dialog Box

Click: [icon] from the Navigator > [Tree Filters...] > check all Display options *on* [Fig. 8.7(c)] > **Apply** > **OK** > **Ctrl+S** > **OK** > **View** tab > [icons] *on* > [icons] *on*

**Figure 8.7(c)** Model Tree Items Dialog Box

Datum planes and the coordinate system are created per the template provided by Creo Parametric 2.0. The datum planes will have the default names, **ASM_RIGHT**, **ASM_TOP**, and **ASM_FRONT**.

Change the coordinate system name: slowly double-click on [ASM_DEF_CSYS] in the Model Tree > *type the new name* [SUB_ASM_CSYS] > **Enter** [Fig. 8.7(d)] > **LMB** in the Graphics Window *(Remember to import your previously created option file from Lesson 6.)*

**Figure 8.7(d)** Sub-Assembly Datum Planes and Coordinate System

Regardless of the design methodology, the assembly datum planes and coordinate system should be on their own separate *assembly layer*. Each part should also be placed on separate assembly layers; the part's datum features should already be on *part layers*. Look over the default template for assembly layering.

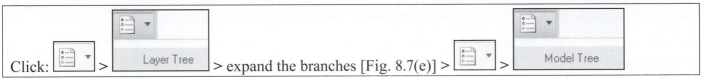

Click: [icon] > Layer Tree > expand the branches [Fig. 8.7(e)] > [icon] > Model Tree

**Figure 8.7(e)** Default Template for Assembly Layering

The first component to be assembled to the subassembly is the Clamp_Arm. The simplest and quickest method of adding a component to an assembly is to match the coordinate systems. The first component assembled is usually where this *constraint* is used, because after the first component is established, few if any of the remaining components are assembled to the assembly coordinate system (with the exception of *top-down design*) or, for that matter, other parts' coordinate systems. Make sure all your models are in the same working directory before you start the assembly process.

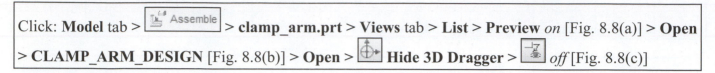

Click: **Model** tab > Assemble > **clamp_arm.prt** > **Views** tab > **List** > **Preview** *on* [Fig. 8.8(a)] > **Open** > **CLAMP_ARM_DESIGN** [Fig. 8.8(b)] > **Open** > Hide 3D Dragger > *off* [Fig. 8.8(c)]

**Figure 8.8(a)** Previewed Clamp_Arm Component

**Figure 8.8(b)** Clamp_Arm_Design    **Figure 8.8(c)** Clamp_Arm_Design in the Graphics Window

When an assembly is complicated, or it is difficult to select constraining geometry, a separate window aids in the selection process. For simple assemblies (as is the Clamp Assembly and Clamp Subassembly) working in the assembly window is more convenient. Also, since this is the first component being added to the assembly, a separate window is not needed.

---

In the Graphics Window, press: **RMB > Default Constraint** [Fig. 8.8(d)] (this option puts the component at the default position on the assembly model, which is the same as using the Coincident constraint) > ☑ > 💾 > **OK > File > Options > Model Display >** Default Model orientation: **Trimetric > Isometric > OK > No >** in the Model Tree, pick: ▶ ( ▼ 🗀 CLAMP_ARM_DESIGN<CLAMP_ARM>.PRT ) > ▶ ( ▼ 🗀 Placement ) > ▶ ( ▼ 🗀 Set1 ) *(your Set # may be different)* ( 🔲 Default ) [Fig. 8.8(e)]

---

**Figure 8.8(d)** Default Constraint

**Figure 8.8(e)** Expanded Placement Folder

335

The Clamp_Arm_Design has been assembled to the default location. This means to align the default Creo Parametric 2.0-created coordinate system of the component to the default Creo Parametric 2.0-created coordinate system of the assembly. Creo Parametric 2.0 places the component at the assembly origin. By using the constraint, Coincident and selecting the assembly and then the component's coordinate systems would have accomplished the same thing, but with more picks.

Component placement is based on placement definition sets. These sets determine how and where the component relates to the assembly. The sets are either user-defined or predefined. A user-defined constraint set has 0 or more constraints (a packaged component may have no constraints). Predefined constraint sets have a predefined number of constraints. Placement of a component in an assembly is determined by the constraints in all sets defined. A single set of constraints can define placement of a component. If constraints from one set conflict with constraints from another set, the placement status becomes invalid. The constraints must be redefined or removed until placement status becomes valid. Constraints can be added or deleted at will in a user-defined constraint set, there are no predefined constraints. Each type of predefined constraint set (also called a connection) has a predefined number of constraints. Constraint sets are displayed in the Placement folder of the Model Tree. Display hierarchy follows the order in which they were defined.

Click: [Assemble] > select **clamp_swivel.prt** from the Open dialog box > **Preview** *(if not already on)* > **Open** > [icon] > **(Select All)** *off* > **LMB** in the Graphics Window > [icon] *on* > **Placement** tab [Fig. 8.9(a)]

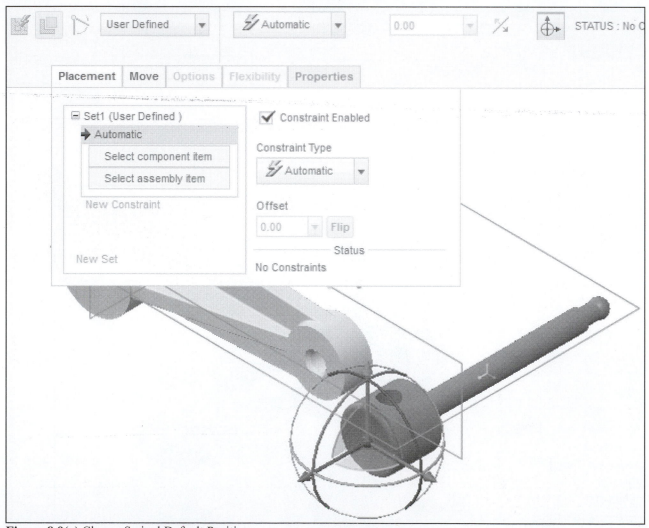

**Figure 8.9(a)** Clamp_Swivel Default Position

Click: **Placement** tab (closed) > pick on the cylindrical surface of the Clamp_Swivel [Fig. 8.9(b)] > pick on the hole surface of the Clamp_Arm_Design [Fig. 8.9(c)] (constraint becomes Coincident) > **View** tab >  **Plane Display** *on* > **Plane Tag Display** *off* > **Component Placement** tab

CLAMP_SWIVEL:Surf:F5(REVOLVE_1)

**Figure 8.9(b)** Select on the Clamp_Swivel Surface

Automatic

CLAMP_ARM_DESIGN:Surf.F23(HOLE_2)

**Figure 8.9(c)** Select on the Clamp_Arm_Design Hole Surface

Place the pointer on the Dragger Axis, press and hold: **LMB** on the Dragger Axis [Fig. 8.9(d)] > pull forward > release the **LMB** > press **RMB** > **New Constraint** [Fig. 8.9(e)]

**Figure 8.9(d)** Pull the Dragger Axis Forward

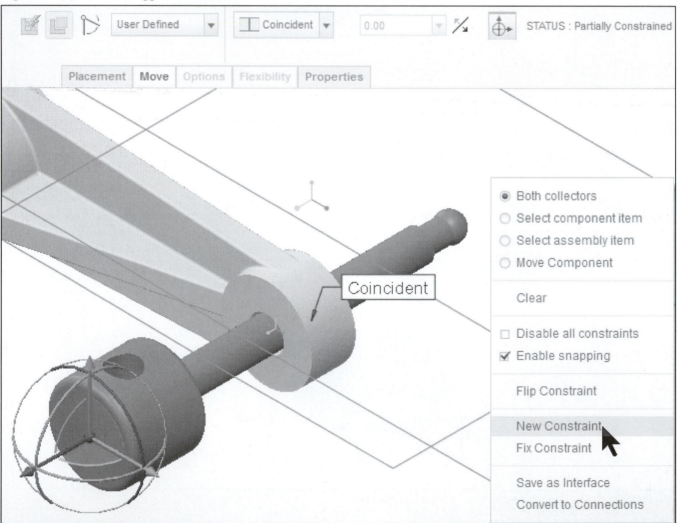

**Figure 8.9(e)** Press RMB > New Constraint

Click:  **Hide 3D Dragger** > **Placement** tab > pick the surface of the Clamp_Arm_Design [Fig. 8.9(f)] > click the **RMB** to toggle to the desired surface > pick the surface of the Clamp_Swivel [Fig. 8.9(g)]

**Figure 8.9(f)** Select on the Top Surface of the Small Circular Extrusion of the Clamp_Arm_Design

**Figure 8.9(g)** RMB > Select on the Underside Surface of the Clamp_Swivel

Click:  **> LMB** on the handle and drag the Clamp_Swivel until it is **1.50** offset from the Clamp_Arm_Design surface [Fig. 8.9(h)] > release the **LMB** > **Placement** tab *off* (closed) [Fig. 8.9(i)]

**Figure 8.9(h)** Drag the handle until **1.50** for the (Mate) Offset Distance

**Figure 8.9(i)** Fully Constrained Component

Click: ✔ [Fig. 8.9(j)] > in the Model Tree, pick: ▸ next to CLAMP_SWIVEL.PRT > ▸ > ▸

( ) *(your Set # may be different)* > 🔍 **Refit** > 💾 > **OK** > **File** > **Manage File** >

Enter object whose old versions should be deleted

CLAMP_SUBASSEMBLY.ASM

**Delete Old Versions** ( ) > ✔

**Figure 8.9(j)** Placement Status

# Regenerating Models

You can use **Regenerate** to find bad geometry, broken parent-child relationships, or any other problem with a part feature or assembly component. In general, it is a good idea to regenerate the model every time you make a change, so that you can see the effects of each change in the graphics window as you build the model. By regenerating often, it helps you stay on course with your original design intent by helping you to resolve failures as they happen.

When Creo Parametric 2.0 regenerates a model, it recreates the model feature by feature, in the order in which each feature was created, and according to the hierarchy of the parent-child relationship between features.

In an assembly, component features are regenerated in the order in which they were created, and then in the order in which each component was added to the assembly. Creo Parametric 2.0 regenerates a model automatically in many cases, including when you open, save, or close a part or assembly or one of its instances, or when you open an instance from within a Family Table. You can also use the Regenerate command to manually regenerate the model.

**Regenerate**, lets you recalculate the model geometry, incorporating any changes made since the last time the model was saved. If no changes have been made, Creo Parametric 2.0 informs you that the model has not changed since the last regeneration. The **Regeneration Manager** opens the Regeneration Manager dialog box if there are features or components that have been changed that require regeneration. A column next to the Regeneration List indicates each entries Status (Regenerated or Unregenerated).

In the dialog box you can:

- Select all features/components for regeneration, select **Show Checked > Check All > Regenerate**.
- Omit all features/components from regeneration, select **Show Checked > Uncheck All > Regenerate**.
- Determine the reason an object requires regeneration, select an entry in the **Regeneration List > RMB > Feature Info**.

Double-click on the Clamp_Swivel > double-click on the **1.50** dimension > *type* **1.75** [Fig. 8.10(a)] > **Enter** >  next to > Regeneration Manager dialog box opens [Fig. 8.10(b)] > **Regenerate** [Fig. 8.10(c)] > **Undo** (dimension is now 1.50) > double-click on the Clamp_Swivel > in the Graphics Window, **LMB** to deselect > **Ctrl+S** > **Enter**

**Figure 8.10(a)** Double-click on the Clamp_Swivel

**Figure 8.10(b)** Modify the Offset Value to **1.75**

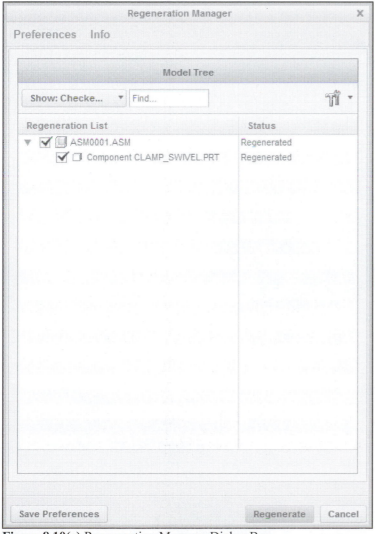

**Figure 8.10(c)** Regeneration Manager Dialog Box

The next component to be assembled is the Clamp_Foot. Click: 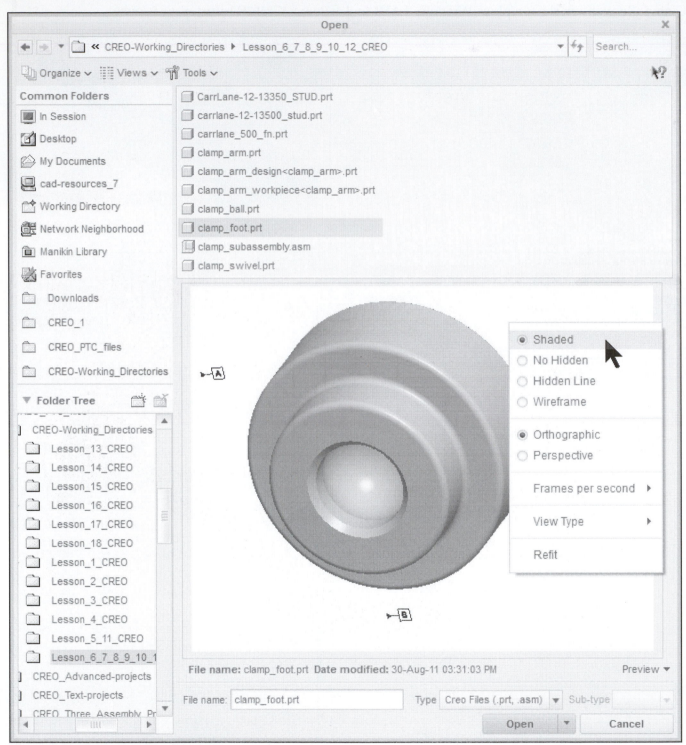 **Assemble** > select the **clamp_foot.prt** from the Open dialog box > **Preview** *on* > **MMB** rotate the model in the Preview Window > **RMB** to see options (Fig 8.11) > **No Hidden** > **RMB** > **Shaded** > **Open**

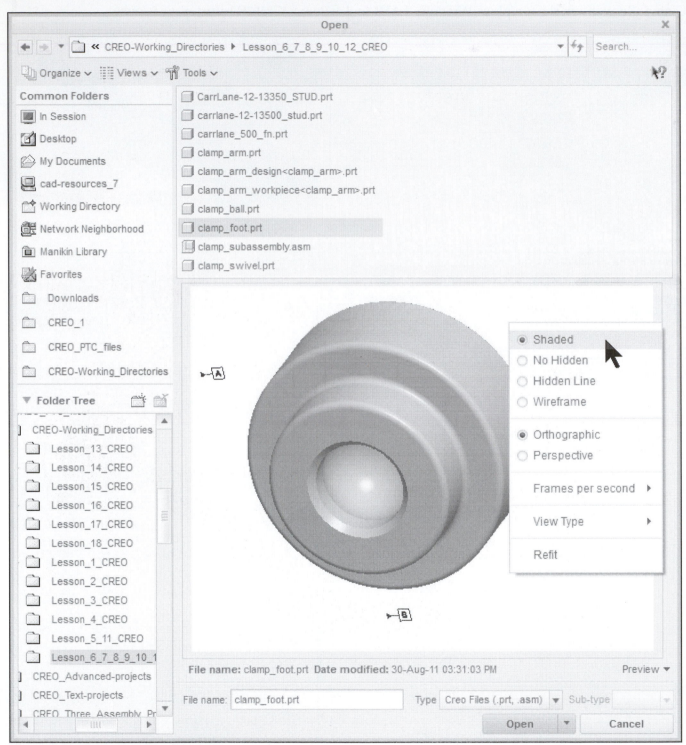

**Figure 8.11** Clamp_Foot Preview

Click:  **Show component in a separate window while specifying constraints** > **Show 3D Dragger** [Fig. 8.12(a)] > use the colored axes draggers to reposition the component model

**Figure 8.12(a)** Component in a Separate Window while Specifying Constraints

Click:  **Hide 3D Dragger** > resize your windows as desired > from the Accessory (Component) Window, pick the *internal* spherical surface of the Clamp_Foot [Fig. 8.12(b)] > pick the *external* spherical surface of the Clamp_Swivel [Fig. 8.12(c)] > `Coincident` from the Dashboard > `Centered`

**Figure 8.12(b)** Select the Internal Spherical Surface of the Clamp_Foot

**Figure 8.12(c)** Select the External Spherical Surface of the Clamp_Swivel

346

From the Accessory (Component) Window, pick the surface of the Clamp_Foot [Fig. 8.12(d)] > from the assembly window, click: **RMB** to toggle to the desired surface > pick the surface of the Clamp_Arm_Design [Fig. 8.12(e)] > **Placement** tab

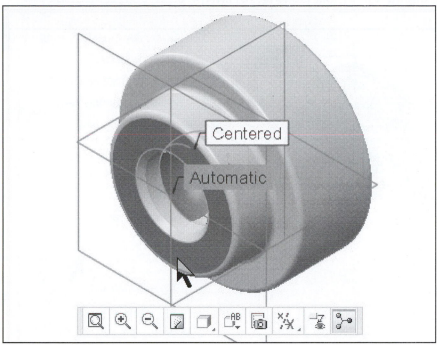

**Figure 8.12(d)** Select the External Cylindrical Surface of the Clamp_Foot

**Figure 8.12(e)** Select the External Cylindrical Surface of the Clamp_Arm_Design

Click:  *off* > `⊤ Coincident` > `∥ Parallel` > from the Graphics Window, press **RMB** [Fig. 8.12(f)] >

**Flip Constraint** [Fig. 8.12(g)] > **Flip** from the Placement tab [Fig. 8.12(h)] > ☑ Allow Assumptions / Fully Constrained

**Figure 8.12(f)** Flip Constraint

**Figure 8.12(g)** Flip Constraint

**Figure 8.12(h)** Original Orientation

Placement tab is still open [Fig. 8.12(i)]. Click: ✔ > 🔲 > | Standard Orientation | > 💾 > **OK**

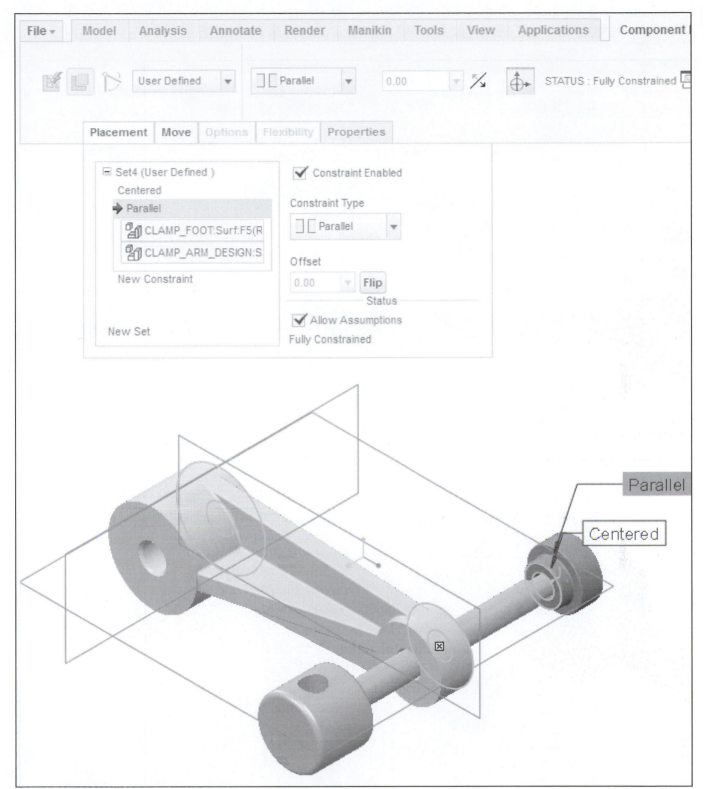

**Figure 8.12(i)** Fully Constrained Component

Assemble the **5.00** stud. Click: [☐ Assemble] > select **Carrlane-12-13500_STUD.prt > Open >** [☐] **Show component in a separate window** *off* > [⊕] **Show 3D Dragger** > rotate the part using the colored arc draggers [Fig. 8.13(a)] (you can rotate and move the component as desired [Fig. 8.13(b)])

**Figure 8.13(a)** Press and hold down the LMB on the Arc

**Figure 8.13(b)** Move the Mouse Pointer to Rotate the Part > at the Desired Orientation, Release the LMB

Click:  **Hide 3D Dragger** > pick the external cylindrical surface of the Carrlane-12-13500_STUD [Fig. 8.13(c)] > pick the internal cylindrical surface of the hole of the Clamp_Swivel [Fig. 8.13(d)]

CARRLANE-12-13500_STUD:Surf:F2(NEUTRAL)

**Figure 8.13(c)** Select the Stud's Cylindrical Surface

CLAMP_SWIVEL:Surf:F6(HOLE_1) Automatic

**Figure 8.13(d)** Select the Swivel's Hole's Cylindrical Surface

351

Pick the end surface of the Carrlane-12-13500_STUD [Fig. 8.14(a)] > pick datum **C** of the Clamp_Swivel [Fig. 8.14(b)]

**Figure 8.14(a)** Select the End Surface

**Figure 8.14(b)** Select the Datum

Click: [icon] > [icon] > drag the handle to **2.50** [Fig. 8.14(c)] > **Placement** tab [Fig. 8.14(d)] > [✓]

**Figure 8.14(c)** Constraint Type changed to Distance *(may already be Distance, depending on the initial location of the stud)*

**Figure 8.14(d)** Press LMB on the Handle and Drag to an Offset of **2.50** > Release the LMB

353

Click: **Ctrl+S > OK >**  **>** ⊞ Tree Columns.. **>** use ≫ to add the columns names to be Displayed [Fig. 8.15(a)] **> Apply > OK >** resize the Model Tree and Model Tree column widths [Fig. 8.15(b)]

**Figure 8.15(a)** Model Tree Columns Dialog Box

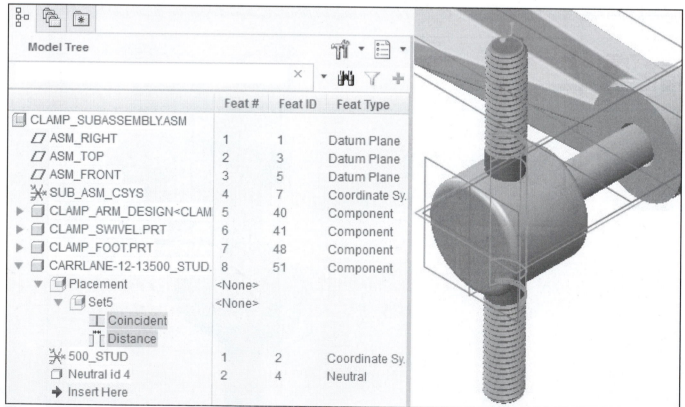

**Figure 8.15(b)** Model Tree with Adjusted Columns

The Clamp_Ball handles are the last components of the Clamp_Subassembly. Reduce the Model Tree size to show only the Feat # column. Click: [Assemble] > select **clamp_ball.prt** > **Preview** *on* > **MMB** spin and zoom the Preview Window [Fig. 8.16(a)] > **Open** > **MMB** to spin the model as shown [Fig. 8.16(b)]

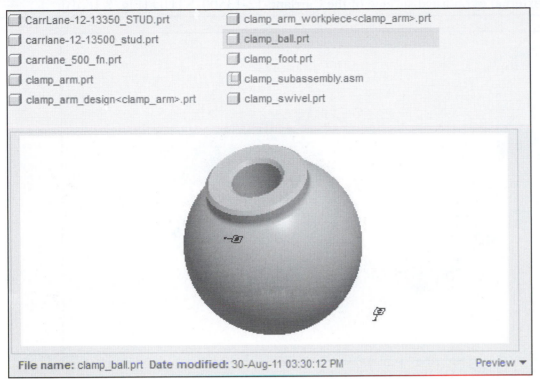

CarrLane-12-13350_STUD.prt      clamp_arm_workpiece<clamp_arm>.prt
carrlane-12-13500_stud.prt       clamp_ball.prt
carrlane_500_fn.prt              clamp_foot.prt
clamp_arm.prt                    clamp_subassembly.asm
clamp_arm_design<clamp_arm>.prt  clamp_swivel.prt

File name: clamp_ball.prt  Date modified: 30-Aug-11 03:30:12 PM        Preview ▼

**Figure 8.16(a)** Clamp_Ball

**Figure 8.16(b)** Spin the Subassembly

To move the new component independently in the assembly window, press and hold **Ctrl+Alt > MMB** on the Clamp_Ball and rotate > **Ctrl+Alt > RMB** pan as needed > release the **RMB** and **Ctrl+Alt** keys >  **Hide 3D Dragger** > pick on the internal cylindrical surface of the hole of the Clamp_Ball [Fig. 8.16(c)] > pick on the external cylindrical surface of the Carrlane-12-13500_STUD [Fig. 8.16(d)]

CLAMP_BALL:Surf:F6(HOLE_1)

**Figure 8.16(c)** Select the Hole's Surface

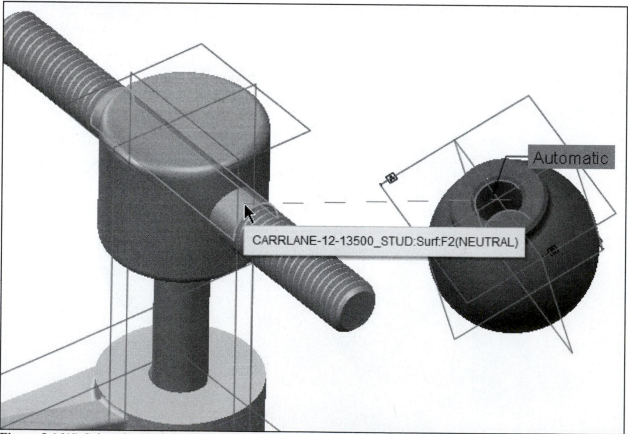

Automatic

CARRLANE-12-13500_STUD:Surf.F2(NEUTRAL)

**Figure 8.16(d)** Select the Shaft Surface of the Carrlane-12-13500_STUD

Click: **Placement** tab in the Dashboard > **New Constraint** > **Ctrl+Alt** > **RMB** pan as needed > pick the end surface of the Carrlane-12-13500_STUD [Fig. 8.16(e)] > pick the flat end of the Clamp_Ball [Fig. 8.16(f)] (**RMB** to toggle to the desired surface if needed)

**Figure 8.16(e)** Select the End Surface of the Carrlane-12-13500_STUD *(your component orientation may appear differently)*

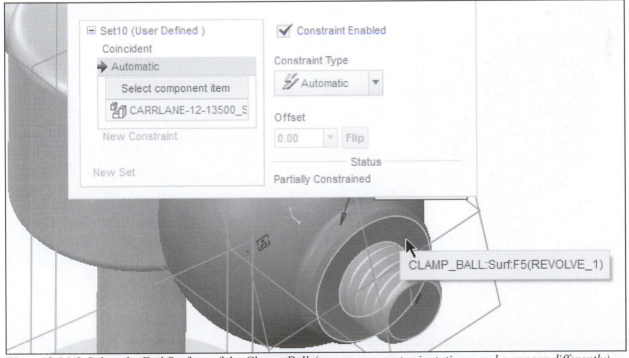

**Figure 8.16(f)** Select the End Surface of the Clamp_Ball *(your component orientation may be appear differently)*

Click: Offset **Flip** (changes the constraint type from Align to Mate) *(Flip as needed)* [Fig. 8.16(g)] > | Coincident | > | Distance | (may already be Distance, based on the initial location of the component) > drag the handle toward the Clamp_Swivel until **.50** [Fig. 8.16(h)] *(or type -.50 > Enter)* > ✓

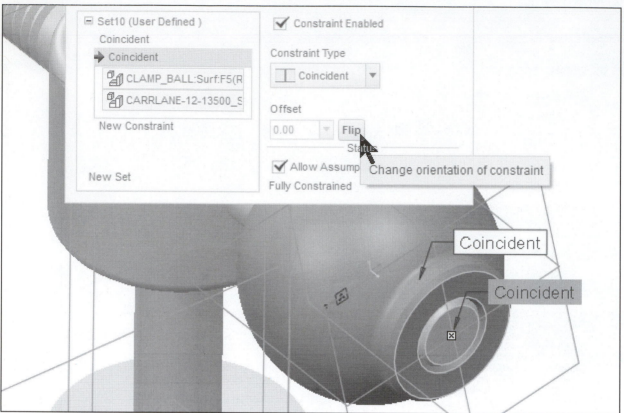

**Figure 8.16(g)** Flip as needed

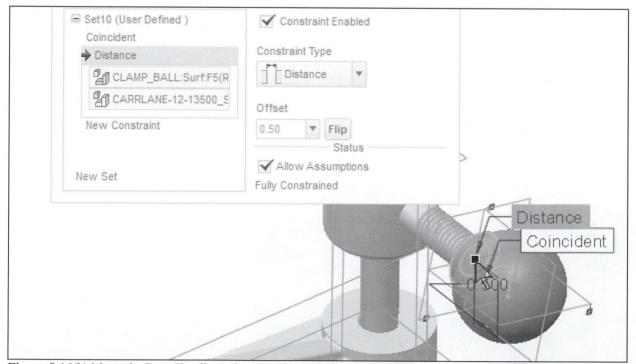

**Figure 8.16(h)** Move the Drag Handle so that the Clamp_Ball is Offset **.500**

Click:  > **OK** > **MMB** to spin the model > pick on the **Clamp_Ball** component to select it *(highlights)* [Fig. 8.16(i)] > ⬜ > ⬛ **Hidden Line** > **Ctrl+C** (copy) > **Ctrl+V** (paste) > pick the opposite end cylindrical surface of the Carrlane-12-13500_STUD as the new reference [Fig. 8.16(j)]

**Figure 8.16(i)** First Clamp_Ball Selected *(highlighted)*

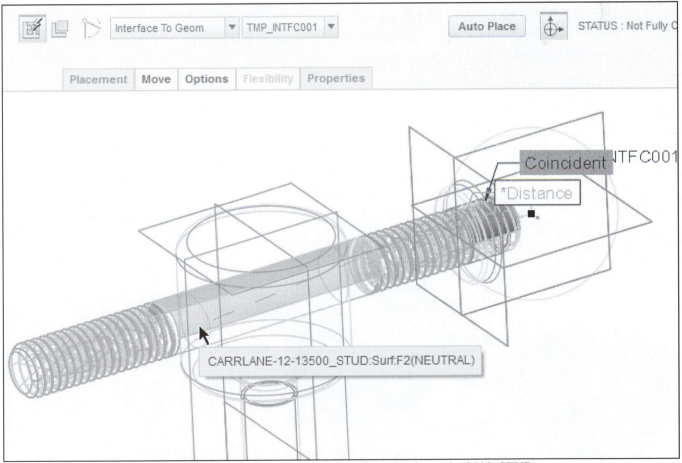

**Figure 8.16(j)** Select the Opposite End of the Cylindrical Surface of the Carrlane-12-13500_STUD

Pick the opposite end (flat) surface of the Carrlane-12-13500_STUD [Fig. 8.16(k)] > ⬚ > ⬚ **Shading** > press **RMB** > **Flip Constraint** > in the Graphics Window; double-click on the **.500** dimension > *type* **.4375** (to modify the distance from the end of the shaft so that it does not bottom-out) [Fig. 8.16(l)] > **Enter** > ✓ > **Ctrl+S** > **OK** > in the Graphics Window, **LMB** to deselect

**Figure 8.16(k)** Select the Opposite End (Flat) Surface of the Carrlane-12-13500_STUD

**Figure 8.16(l)** Offset **.4375**

Double-click on the first Clamp_Ball and modify the offset distance from **.500** to **.4375** > **Enter** >

**Regenerate** > **Ctrl+D** > 🔍 **Command Search** > *type* **bill** > **Bill of Materials** [Fig. 8.17(a)] > ⦿ Top Level [Fig. 8.17(b)] > **OK** [Fig. 8.17(c)] > close the Browser

**Figure 8.17(a)** Command Search: Bill of Materials

**Figure 8.17(b)** BOM Dialog Box

## Bom Report : CLAMP_SUBASSEMBLY

**Assembly CLAMP_SUBASSEMBLY contains:**

| Quantity | Type | Name | Actions |
|---|---|---|---|
| 1 | Part | CLAMP_ARM_DESIGN | |
| 1 | Part | CLAMP_SWIVEL | |
| 1 | Part | CLAMP_FOOT | |
| 1 | Part | CARRLANE-12-13500_STUD | |
| 2 | Part | CLAMP_BALL | |

**Summary of parts for assembly CLAMP_SUBASSEMBLY:**

| Quantity | Type | Name | Actions |
|---|---|---|---|
| 1 | Part | CLAMP_ARM_DESIGN | |
| 1 | Part | CLAMP_SWIVEL | |
| 1 | Part | CLAMP_FOOT | |
| 1 | Part | CARRLANE-12-13500_STUD | |
| 2 | Part | CLAMP_BALL | |

**Figure 8.17(c)** Clamp_Subassembly Bill of Materials (BOM)

Click: **View** tab > [icons] *on* > [icons] *on* > [icon] **View Manager** from the Status Bar > **Sections** tab [Fig. 8.18(a)] > **New** > **Planar** [Fig. 8.18(b)] > *type* **A** [Fig. 8.18(c)] > **Enter** > pick datum **ASM_TOP** [Fig. 8.18(d)] > [checkmark] [Fig. 8.18(e)]

**Figure 8.18(a)** Sections

**Figure 8.18(b)** New > Planar

**Figure 8.18(c)** A is the New Cross Section Name

**Figure 8.18(d)** Select Assembly Datum Top

**Figure 8.18(e)** Section A

362

Click: [icons] *off* > [→A] > **RMB** > ☑ Show Section [Fig. 8.18(f)] > No Cross Section > **RMB** > **Activate** [Fig. 8.18(g)] > [👁 A] > **RMB** > ☐ Show Section > **Close > Ctrl+D > Ctrl+S > OK > File > Manage File > Delete Old Versions > MMB > File > Close** (close the subassembly)

**Figure 8.18(f)** Show Section A Cross Hatching

**Figure 8.18(g)** No Cross Section Set Active

# Part Two: Top-Down Design

## Swing Clamp Assembly

The first features for the main assembly will be the default datum planes and coordinate system. The first part assembled on the main assembly will be the Clamp_Plate, which will be created using *top-down design*; where the assembly is active, and the component is created within the assembly mode. The subassembly is still *"in session-in memory"* even though it does not show on the screen after its window is closed. For the two standard parts of the main assembly, you will use specific constraints instead of using Automatic, which allows Creo Parametric 2.0 to default to an assumed constraint.

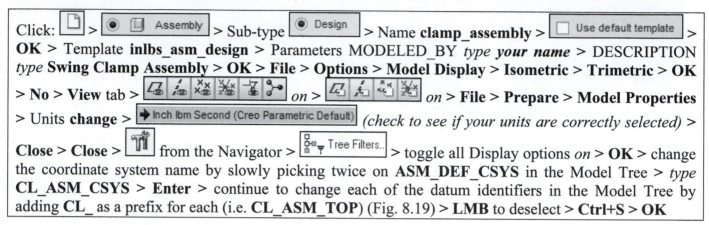

Click: [ ] > (●) [ ] Assembly > Sub-type (●) Design > Name **clamp_assembly** > [ ] Use default template > **OK** > Template **inlbs_asm_design** > Parameters MODELED_BY *type your name* > DESCRIPTION *type* **Swing Clamp Assembly** > **OK** > **File** > **Options** > **Model Display** > **Isometric** > **Trimetric** > **OK** > **No** > **View** tab > [icons] *on* > [icons] *on* > **File** > **Prepare** > **Model Properties** > **Units change** > [ Inch lbm Second (Creo Parametric Default) ] *(check to see if your units are correctly selected)* > **Close** > **Close** > [ ] from the Navigator > [ Tree Filters.. ] > toggle all Display options *on* > **OK** > change the coordinate system name by slowly picking twice on **ASM_DEF_CSYS** in the Model Tree > *type* **CL_ASM_CSYS** > **Enter** > continue to change each of the datum identifiers in the Model Tree by adding **CL_** as a prefix for each (i.e. **CL_ASM_TOP**) (Fig. 8.19) > **LMB** to deselect > **Ctrl+S** > **OK**

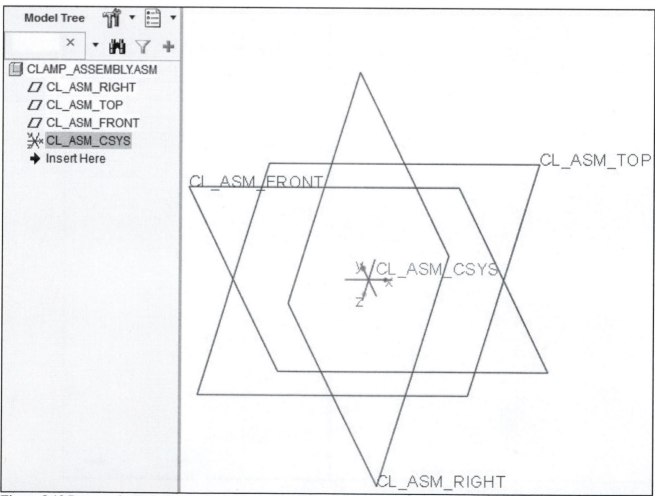

**Figure 8.19** Rename the Default Coordinate System and Datum Planes

## Creating Components in the Assembly Mode

The Clamp_Subassembly is now complete. The subassembly will be added to the assembly after the Clamp_Plate is created and assembled. Using [Create] and the Component Create dialog box (Fig. 8.20) you can create different types of components: parts, subassemblies, skeleton models, and bulk items. You cannot reroute components created in the Assembly mode.

The following methods allow component creation in the context of an assembly without requiring external dependencies on the assembly geometry:

- Create a component by copying another component or existing start part or start assembly
- Create a component with default datums
- Create an empty component
- Create the first feature of a new part; this initial feature is dependent on the assembly
- Create a part from an intersection of existing components
- Create a mirror copy of an existing part or subassembly
- Create Solid or Sheetmetal components
- Mirror Components

**Figure 8.20** Component Create Dialog Box

## Main Assembly, Top-Down Design

The Clamp_Plate component is the first component of the main assembly. The Clamp_Plate is a new part. You will be modeling the plate *"inside"* the assembly using *top-down design*. The drawing in Figure 8.21 provides the dimensions necessary to model the Clamp_Plate.

**Figure 8.21** Clamp_Plate Detail Drawing

Click: **Model** tab > 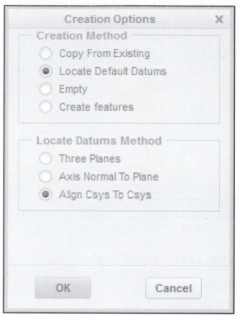 **Create a component in assembly mode** Component Create dialog box displays > Type ⦿ Part > Sub-type ⦿ Solid > Name **clamp_plate** [Fig. 8.22(a)] > **OK** Creation Options dialog box displays [Fig. 8.22(b)] > ⦿ Locate Default Datums > ⦿ Align Csys To Csys > **OK** > pick **CL_ASM_CSYS** from the Graphics Window, ▢ CLAMP_PLATE.PRT displays in the Model Tree, *the small symbol/icon* ▢ *indicates that this component is now active* > expand and highlight the items in the Model Tree for the CLAMP_PLATE.PRT [Fig. 8.23(a)] > **LMB** in the Graphics Window to deselect

**Figure 8.22(a)** Component Create Dialog Box

**Figure 8.22(b)** Creation Options Dialog Box

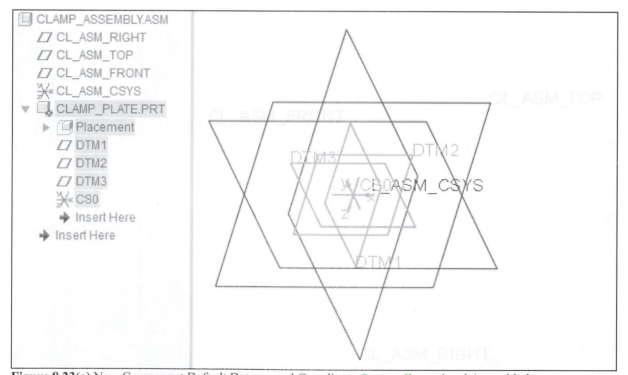

**Figure 8.23(a)** New Component Default Datums and Coordinate System Created and Assembled

366

Assembly Tools are now *unavailable* in the Model Ribbon. The current component is the Clamp_Plate. *You are now effectively in Part mode*, except that you can see the assembly features and components.

Create the protrusion, click: [Extrude] > change the depth value to **.750** [0.7500] > **Enter > Placement** tab > **Define** [Sketch dialog box] Sketch dialog box opens > Sketch Plane--- Plane: select **DTM3** [Fig. 8.23(b)] > **Sketch** [Fig. 8.23(c)]

**Figure 8.23(b)** Sketch Plane and Reference Orientation

**Figure 8.23(c)** Sketch Dialog Box

Click: [icon] > in the Graphics Window, press **RMB > Construction Centerline** > sketch a *vertical* centerline > sketch a *horizontal* centerline > move the mouse off of the centerline > **MMB > LMB >**

[icon] **Create Corner Rectangle** > pick the two corners of the rectangle > **MMB** > LMB *(if you selected the corners carefully you will get a square with only one dimension, if not, click:* [=] *Equal > pick one horizontal line and one vertical line > MMB > MMB)* [Fig. 8.23(d)] > double-click on the remaining dimension > *type* **3.00** [Fig. 8.23(e)] > **Enter > LMB** to deselect > [✓]

**Figure 8.23(d)** Sketch a Square Section

**Figure 8.23(e)** Modify the Value to **3.00**

Press: **Ctrl+D** [Fig. 8.23(f)] > ✓ > ⬚ > ⬚ **Hidden Line** [Fig. 8.23(g)] > pick on the

**CLAMP_ASSEMBLY.ASM** in the Model Tree > **RMB** > **Activate** > **Ctrl+S** > **OK** > pick on the **CLAMP_PLATE.PRT** in the Model Tree > **RMB** > **Activate** (again, the Clamp_Plate is now active ⬦ CLAMP_PLATE.PRT ) > **LMB** in the Graphics Window

**Figure 8.23(f)** Depth and Direction Preview

Figure 8.23(g) Completed Protrusion

Throughout the lesson, follow the steps as given. Drag the handles to the appropriate references and let Creo Parametric 2.0 select the references automatically *even if you think the selections are incorrect*. Later, we will address these inconsistencies, but if you correct them on your own, steps described later will not work (or be needed).

Press: **Ctrl** key and select the three assembly datum planes in the Model Tree > release the **Ctrl** key > **RMB** > **Hide** > [Hole] > [Create standard hole] > [toggle *off*] > [ ] > [ ] > [Adds tapping *on*] > select the tap and drill size **1/2-13**
[ ] > **Shape** tab > [Thru Thread] [Fig. 8.23(h)] > **Properties** tab [Fig. 8.23(i)] > Pick on the surface of the protrusion as the placement plane. A preview hole displays with drag handles for position and size [Fig. 8.23(j)].

Figure 8.23(h) Hole Shape

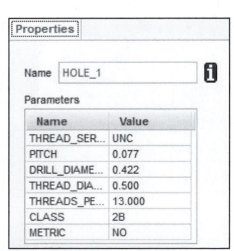

Figure 8.23(i) Hole Properties

369

To establish the Offset References, move one diamond drag handle to the vertical datum (**DTM1**) and the other diamond drag handle to the horizontal datum (**DTM2**) respectively > **Placement** tab > change *both*

| DTM1:F1(DAT... | Align |
| DTM2:F2(DAT... | Align |

linear dimensions to **Align** [Fig. 8.23(k)] > ✓ > **LMB** in the Graphics

▼ ⊔ Hole 1
    A≡ Note_0

Window > **Ctrl+D** > [Fig. 8.23(l)] > 🖫 > **OK** > **LMB** to deselect > press the **Ctrl** key > select the three assembly datum planes in the Model Tree > release the **Ctrl** key > **RMB** > **Unhide** > in the Graphics Window, **LMB** to deselect

**Figure 8.23(j)** Initial Hole Placement (at point of selection) and with Drag Handles Moved

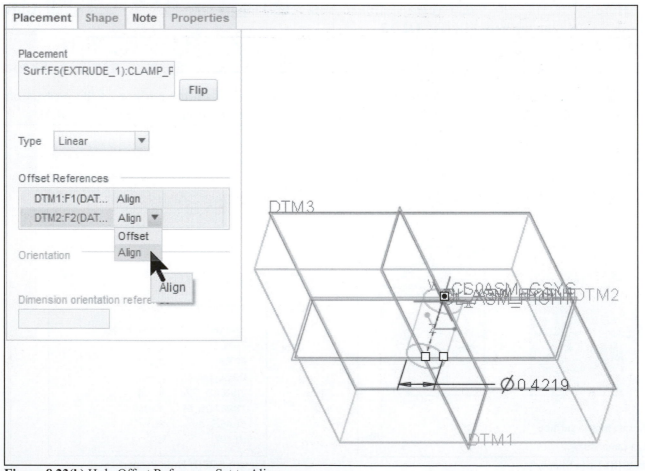

**Figure 8.23(k)** Hole Offset References Set to Align

370

Figure 8.23(l) Completed Tapped Hole

Create a counterbore hole, click:  **Hole** > on > off > on > 5/8-11 > off > on > > **Drill to intersect with all surfaces** > **Shape** tab > input the counterbore values [Fig. 8.23(m)] *(the detail drawing callouts are slightly different)* [Fig. 8.23(n)]

Figure 8.23(m) Counterbore Specifications

Figure 8.23(n) Hole Callout on the Detail Drawing

371

Pick on the front surface to place the counterbore hole [Fig. 8.23(o)] > press **RMB** > **Offset References Collector** > **Placement** tab > pick on **CL_ASM_RIGHT** from the model or Model Tree > hold down the **Ctrl** key and pick on **CL_ASM_TOP** from the model or the Model Tree [Fig 8.23(o)] > release the **Ctrl** key > modify both linear dimensions to **.875** [Fig. 8.23(p)] > ✓ > **LMB** in the Graphics Window

**Figure 8.23(o)** Offset References- Select from the Model or the Model Tree

**Figure 8.23(p) .875** Offset Dimensions *(you may need to use negative .875)*

The holes "seem" correct, but it is important to check for external references. Pick on **Hole 1** from the Model Tree > **RMB** > **Info** > **Reference Viewer** > **Reference Filters** tab > ⦿ References > ☑ Components in path [Fig. 8.23(q)] [DTM1 and DTM2 are Internal (Local) References] > **Close** > pick on **Hole 2** from the Model Tree > **RMB** > **Info** > **Reference Viewer** > **Reference Filters** tab > ⊗ expand [Fig. 8.23(r)] (CL_ASM_RIGHT and CL_ASM_TOP are External References) > **Close**

**Figure 8.23(q)** DTM1 and DTM2 as the HOLE 1 Internal References *(place pointer on top of a leader line to view added info)*

**Figure 8.23(r)** Expand the References to see the CL_ASM_RIGHT and CL_ASM_TOP External References

If you look back at Figure 8.23(o) you will see that the assembly datums were used to place the Hole 2 instead of DTM1 and DTM2. The following commands will show you how to edit the references. If your references are correct- do not change them. If both of your holes have external references, then use the following steps to reference the Hole 2 from DTM1 and DTM2.

Press: **RMB** on the **Hole 2** in the Model Tree > **Edit Definition** [Fig. 8.23(s)]

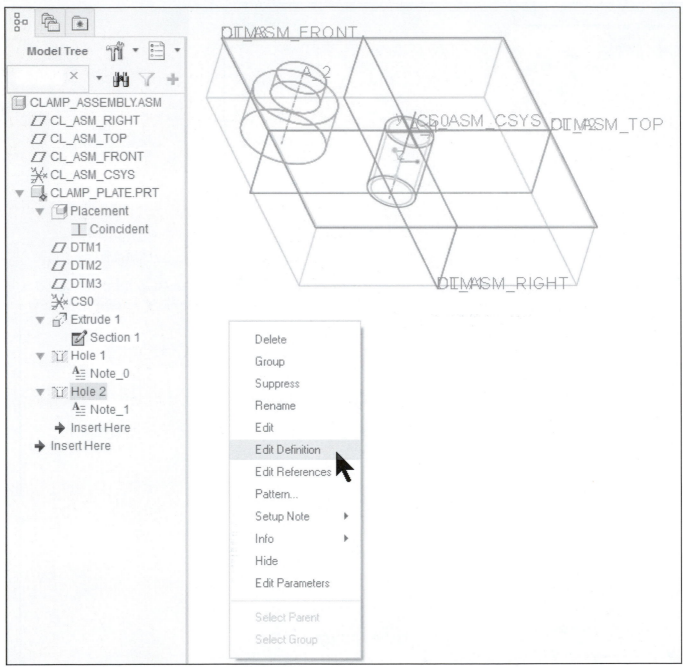

**Figure 8.23(s)** Edit Definition

Click: **Placement** tab > click in the Offset References field > **RMB** > **Remove All** [Fig. 8.23(t)] > place the cursor over datum CL_ASM_TOP to highlight > click **RMB** *[do not hold down the RMB, unless you want to "Pick From List" (accessing the query select list) or choose one of the other options]* and DTM2 highlights [Fig. 8.23(u)] > pick on **DTM2**

**Figure 8.23(t)** Remove All Offset References

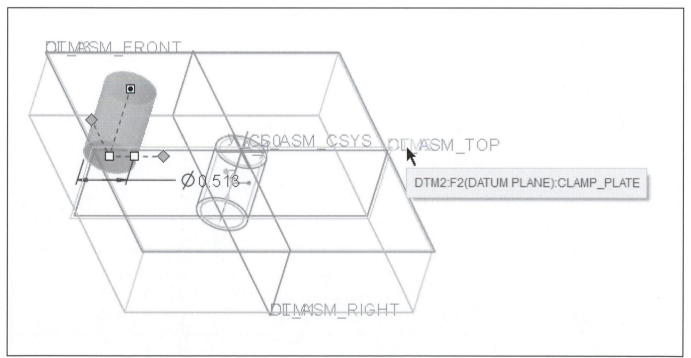

**Figure 8.23(u)** Place Pointer on CL_ASM_TOP > RMB > select DTM2

Press and hold the **Ctrl** key and place the cursor over CL_ASM_RIGHT (highlights) > click **RMB** (DTM1 highlights) [Fig. 8.23(v)] > pick on **DTM1** [Fig. 8.23(w)] *(both datums could also have been selected from the Model Tree)* > release the **Ctrl** key > **.875** is still the Offset distance for both References > **Enter** to complete the command > **Ctrl+S** > **MMB** > in the Graphics Window, **LMB** to deselect

**Figure 8.23(v)** Place Pointer on CL_ASM_RIGHT > RMB > select DTM1

**Figure 8.23(w)** New Offset References

Pick on **Hole 2** from the Model Tree or model > **Tools** tab > **Reference Viewer** > **Reference Filters** tab > **References** *on* >  expand the references [Fig. 8.23(x)] > **Close** > **LMB** to deselect > **Model** tab [If necessary, repeat the previous process to select the correct references (DTM2 and DTM1) for the tapped hole in the center of the part [Fig. 8.23(y)]. The hole will be aligned with both references.]

**Figure 8.23(x)** Reference Viewer (Holes are now Offset from the Datum Planes of the Clamp_Plate)

**Figure 8.23(y)** DTM1 and DTM2 are Offset References for the Tapped Hole

377

Press: **Ctrl+D** > *(if the CLAMP_PLATE.PRT is not active, click on **CLAMP_PLATE.PRT** in the Model Tree > **RMB** > **Activate**)* > double-click on the counterbore hole > double-click on the counterbore diameter dimension > *type* **.8125** > **Enter** > move the mouse > **LMB** > move the mouse > **LMB** > **Ctrl+G** to Regenerate > pick on the counterbore hole in the Model Tree > **RMB** > **Pattern** [Fig. 8.24(a)] > **Dimensions** tab > pick on the *horizontal* **.875** dimension > *type* **–1.75** [Fig. 8.24(b)] > **Enter**

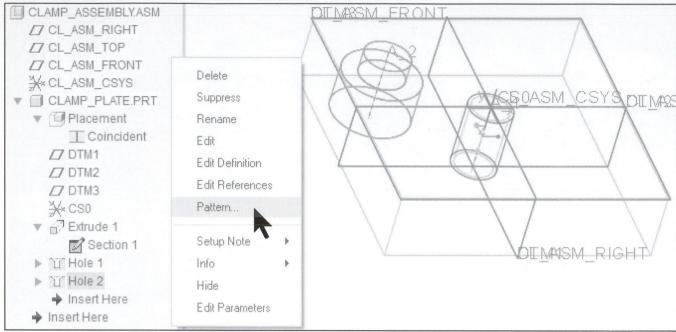

**Figure 8.24(a)** Pattern the Counterbore Hole

**Figure 8.24(b)** Direction 1 Dimension –1.75, Two Items

378

Press: **RMB** > **Direction 2 Dimensions** *(or click in Direction 2 collector box, pick **Select Items**)* > pick on the vertical **.875** dimension [Fig. 8.24(c)] > A "combo box" opens in the Graphics Window, with the dimension increment initially equal to the dimension value. Highlight the value (if needed) and *type* **–1.75** [Fig. 8.24(d)] > **Enter** > ✓ > 🗐 > 💾 > **OK** > **LMB** to deselect

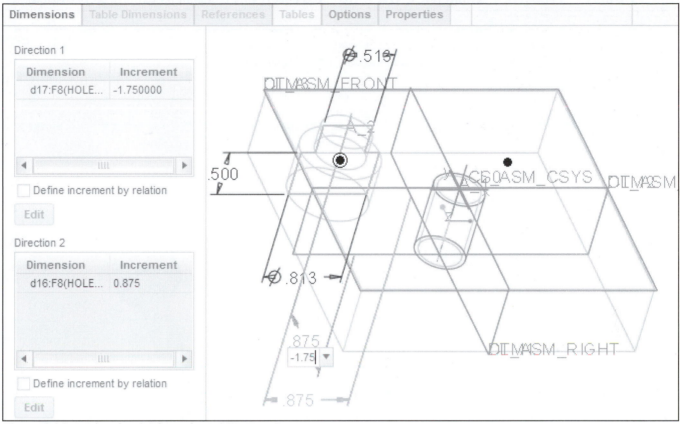

**Figure 8.24(c)** Direction 2 Dimension –1.75, Two Items

**Figure 8.24(d)** Patterned Counterbore Holes

Click: **CLAMP_ASSEMBLY.ASM** from the Model Tree > **RMB** > **Activate** > **View** tab >  *off* >  *off* >  *on* >  *on* >  *on* >  >  > **Model** tab >  **Assemble** > select **clamp_subassembly.asm** > **Preview** *on* [Fig. 8.25(a)] > **Open** > **File** > **Options** > **Model Display** > **Trimetric** > **Isometric** > **OK** > **No** > **Ctrl+D** *(Follow the steps exactly as provided)*

**Figure 8.25(a)** Preview Clamp_Subassembly

380

Click:  **Hide 3D Dragger** > | Automatic | > | Coincident [Fig. 8.25(b)] > **MMB** spin the model > pick the bottom surface of the Clamp_Arm_Design (of the Clamp_Subassembly) [Fig. 8.25(c)]

**Figure 8.25(b)** Assembling the Clamp_Subassembly

**Figure 8.25(c)** Bottom Surface of the Clamp_Arm_Design is selected as the Component Reference

Click: **Placement** tab > pick the non-counterbore surface of the Clamp_Plate [Figs. 8.25(d-e)] > **Flip**

**Figure 8.25(d)** Non-counterbore Surface of the Clamp_Plate is highlighted as the Assembly Reference

**Figure 8.25(e)** Coincident

Click:  > ☑ 🖉 Axis Display > **LMB** in the Graphics Window > **MMB** spin the model > press **RMB** > **New Constraint** > 🖉 > ⊥ **Coincident** > pick the hole axis of the Clamp_Arm_Design [Fig. 8.25(f)] > pick the center tapped hole axis of the Clamp_Plate [Fig. 8.25(g)]

*(Note that you could have picked the two respective hole surfaces to achieve the same result.)*

**Figure 8.25(f)** Select the Axis on the Clamp_Arm_Design

**Figure 8.25(g)** Select the Tapped Hole Axis on the Clamp_Plate

Click: ☑ [Fig. 8.25(h)] > **Ctrl+D** > 🖫 > **OK** > select **CLAMP_PLATE.PRT** [Fig. 8.25(i)] > **LMB**

**Figure 8.25(h)** Fully Constrained Clamp_Subassembly

**Figure 8.25(i)** Assembled Clamp_Subassembly

Click:  > Layer Tree > pick on Layers > **RMB** > **Hide** > Model Tree > > **Shading With Edges** > Assemble > select **Carrlane-12-13350_STUD.prt** > **Preview** *on* [Fig. 8.26(a)] > **Open** > rotate the assembly model using the **MMB** > Show 3D Dragger (if needed) > rotate the short stud (component) part using the colored arc draggers > reposition the component using the center ball of the 3D Dragger [Fig. 8.26(b)] > Hide 3D Dragger

**Figure 8.26(a)** Carrlane-12-13350_STUD.prt

**Figure 8.26(b)** Rotate the Assembly Model and Reposition the Short Stud using the 3D Dragger

Pick on the cylindrical surface of the Carrlane-12-13350_STUD [Fig. 8.26(c)] > pick on the hole surface of the Clamp_Arm_Design [Fig. 8.26(d)]

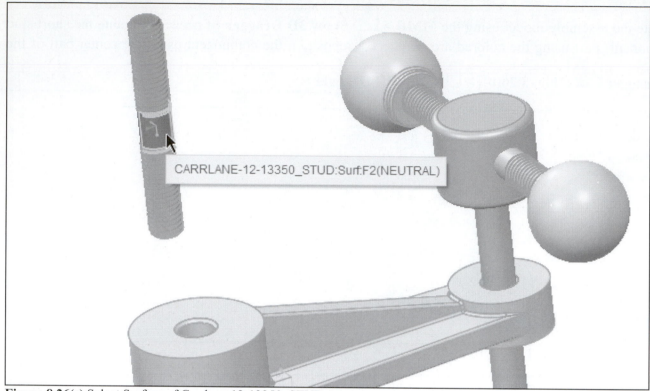

**Figure 8.26(c)** Select Surface of Carrlane-12-13350_STUD.prt

**Figure 8.26(d)** Select Hole Surface of the Clamp_Arm_Design

Press: **Ctrl+D** > **Move** tab > **LMB** on the Carrlane-12-13350_STUD > release the **LMB** > move stud into the hole > **LMB** to place [Fig. 8.26(e)] > 🔲 > **TOP** > 🔲 > 🔲 **Hidden Line** > **LMB** on the Carrlane-12-13350_STUD > translate (slide) until it is inside the Clamp_Plate [Fig. 8.26(f)] > **LMB** to place

**Figure 8.26(e)** Move Tab, Motion Type Translate

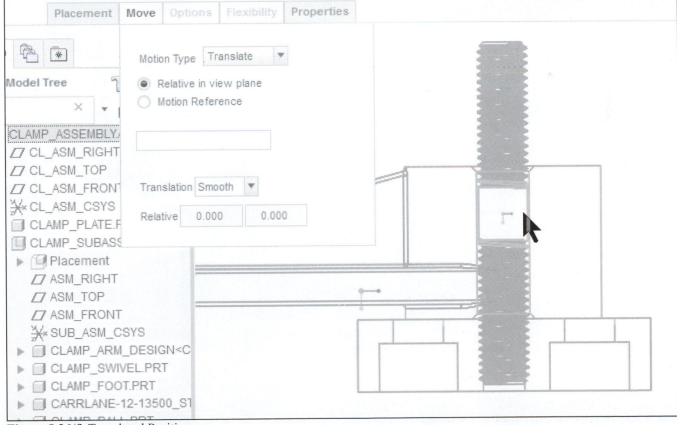

**Figure 8.26(f)** Translated Position

Click: 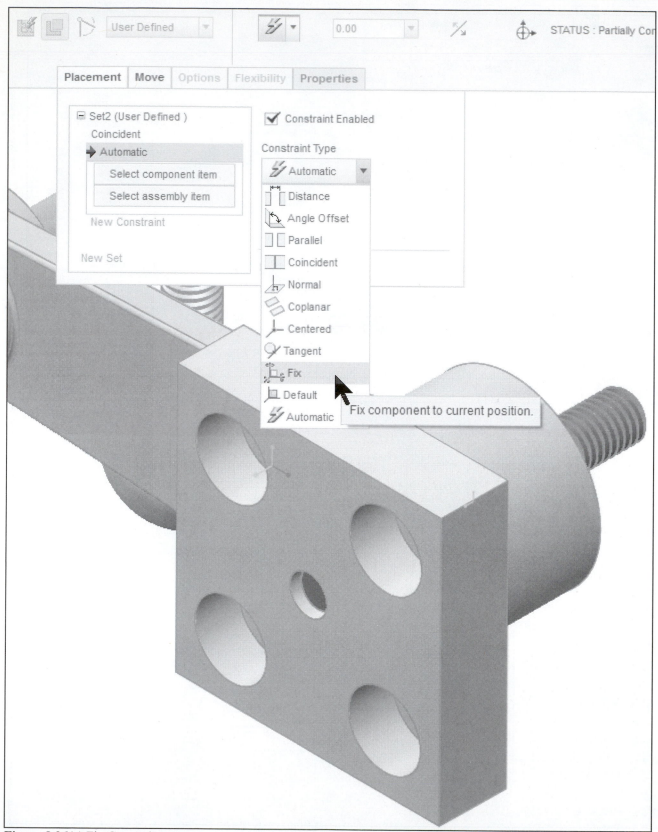 > ▢ > **Ctrl+D** > **Placement** tab > **New Constraint** > Constraint Type ▾ > 🔒 Fix **Fix component to current position** [Fig. 8.26(g)] > ✓ > **Ctrl+D** > **Ctrl+S** > **OK**

**Figure 8.26(g)** Fix Constraint

388

Click:  Assemble > select **carrlane_500_fn.prt** [Fig. 8.27(a)] > **Open** > press and hold **Ctrl+Alt** > press **RMB** on the flange nut and move it closer to the assembly [Fig. 8.27(b)] > release the **RMB**

CarrLane-12-13350_STUD.prt          clamp_arm_workpiece<clamp_arm>.prt
carrlane-12-13500_stud.prt          clamp_assembly.asm
carrlane_500_fn.prt                 clamp_ball.prt
clamp_arm.prt                       clamp_foot.prt
clamp_arm_design<clamp_arm>.prt     clamp_plate.prt

**Figure 8.27(a)** carrlane_500_fn.prt

**Figure 8.27(b)** Move the CARRLANE_500_FN Component

Press and hold **Ctrl+Alt** > press **MMB** on the Flange_Nut and rotate [Fig. 8.27(c)] > release the **MMB** and the **Ctrl+Alt** keys >  **Hide 3D Dragger** (if needed) > in the Graphics Window, press **RMB** > **Select component item** [Fig. 8.27(d)]

**Figure 8.27(c)** Rotate the Flange Nut

**Figure 8.27(d)** Press RMB > Select component item

Carefully pick the cylindrical surface of the flange nut [Fig. 8.27(e)] > carefully pick the cylindrical surface of the stud [Fig. 8.27(f)] (Coincident) *[press **RMB** > **Clear** if Tangent becomes the constraint]*

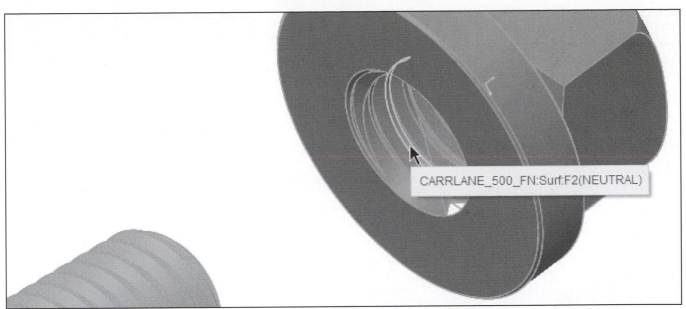

**Figure 8.27(e)** Select the Flange Nut Surface (zoom in if needed)

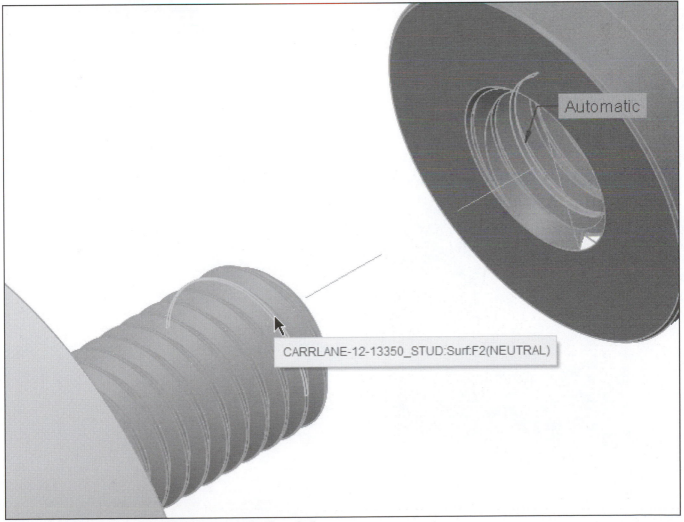

**Figure 8.27(f)** Select the Stud Surface (zoom in if needed)

Press: **MMB** spin the model > **Placement** tab > **New Constraint** > rotate and translate the component as required > pick on the Clamp_Arm_Design surface [Fig. 8.27(g)]

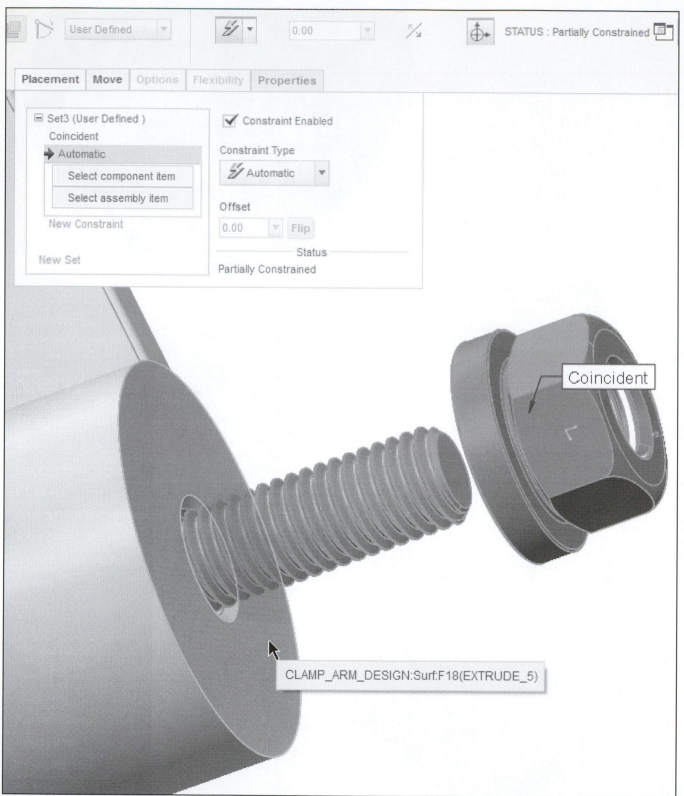

**Figure 8.27(g)** Select the Clamp_Arm_Design Surface

**Figure 8.27(h)** Select the CARRLANE_500_FN Bearing Surface

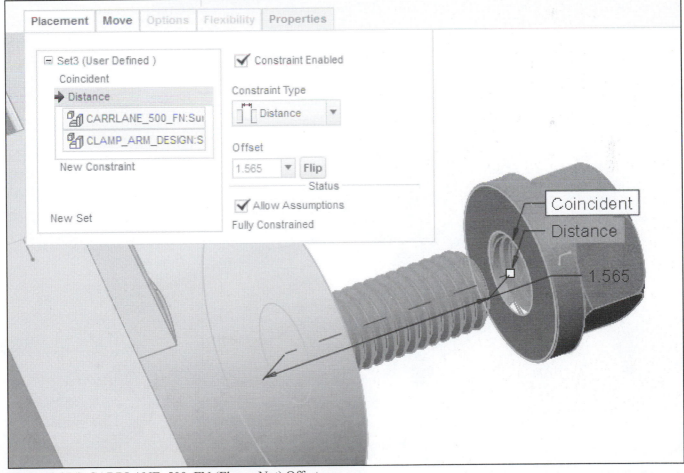

**Figure 8.27(i)** CARRLANE_500_FN (Flange Nut) Offset

**Figure 8.27(j)** Constraint Type: Coincident

Perform a check on the assembly. Click: **Analysis** tab > [ Global Interference ] in the Inspect Geometry Group > [ 👓 ] **Compute current analysis for preview** (Fig. 8.28) *(Select each line and note the interference between components at each threaded connection.)* > ✓ > **Ctrl+S > Enter**

**Figure 8.28** Global Interference Dialog Box

## Bill of Materials

The Bill of Materials (BOM) lists all parts and part parameters in the current assembly or assembly drawing. It can be displayed in HTML or text format and is separated into two parts: breakdown and summary. The Breakdown section lists what is contained in the current assembly or part. The BOM HTML breakdown section lists quantity, type, name (hyperlink), and three actions (highlight, information and open) about each member or sub-member of your assembly:

* **Quantity** Lists the number of components or drawings
* **Type** Lists the type of the assembly component (part or sub-assembly)
* **Name** Lists the assembly component and is hyper-linked to that item. Selecting this hyperlink highlights the component in the graphics window
* **Actions** is divided into three areas:

  o **Highlight** Highlights the selected component in the assembly graphics window

  o **Information** Provides model information on the relevant component

  o **Open** Opens the component in another Creo Parametric 2.0 window

A bill of materials (BOM) can be seen by clicking: **Tools** tab > [Bill of Materials] > check all Include options [Fig. 8.29(a)] > **OK** [Fig. 8.29(b)] > [×] or [⊘] to close the Browser > **Model** tab

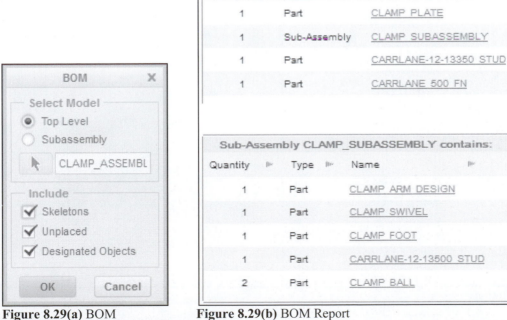

Figure 8.29(a) BOM          Figure 8.29(b) BOM Report

Upon closer inspection, the assembly is not assembled correctly. The Clamp_Subassembly is assembled on the wrong side of the Clamp_Plate. When using Creo Parametric 2.0, you can edit items, features, components, etc., at any time. The Edit Definition command is used to change the constraint reference of assembled components.

*(If necessary, Open the Clamp_Assembly.)* In the Quick Access Toolbar, click:  >
◉ 1 CLAMP_ASSEMBLY.ASM *[or from the **View** tab (Fig. 8.30)]* to activate the top assembly model > select the **CLAMP_SUBASSEMBLY.ASM** in the Model Tree > **RMB** > **Edit Definition** [Fig. 8.31(a)]

**Figure 8.30** Activate the Assembly

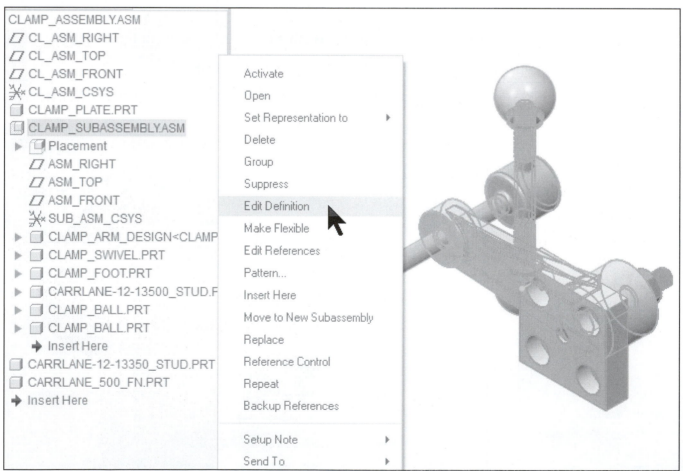

**Figure 8.31(a)** RMB > Edit Definition

Click: **Placement** tab (to see the existing references) > pick the second (assembly) reference
CLAMP_PLATE:Surf:F5(I  from the Placement dialog box [Fig. 8.31(b)] *(check the Tool Tip*
Assembly Ref: CLAMP_PLATE:Surf:F5(EXTRUDE_1) *to make sure you selected the Clamp_Plate reference)*

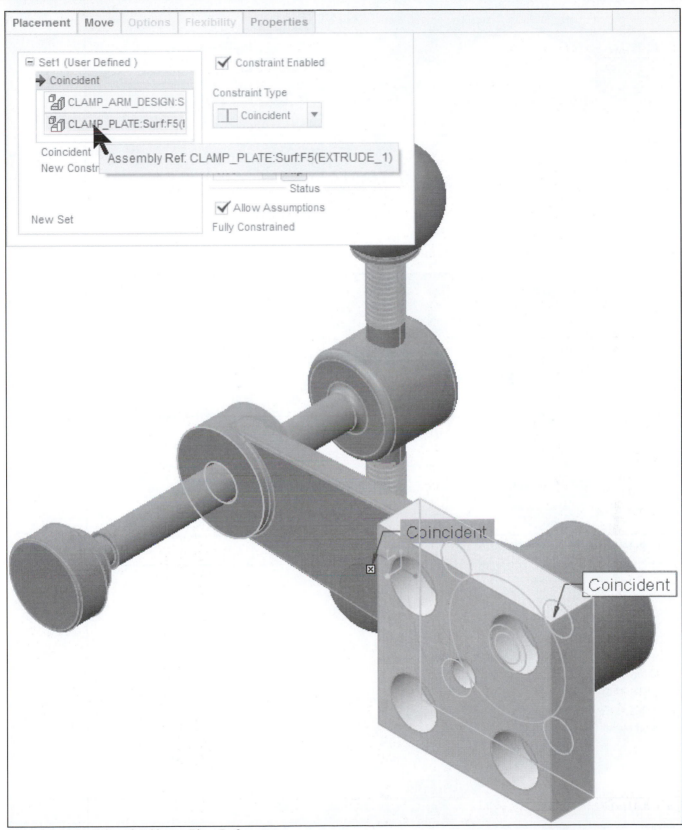

**Figure 8.31(b)** Select the Clamp_Plate Reference

Pick the counterbore side of the Clamp_Plate as the new assembly reference [Fig. 8.31(c)] > in the Graphics Window, press **RMB** > **Flip Constraint** [Fig. 8.31(d)] > [✓] > **Ctrl+D** [Fig. 8.31(e)]

**Figure 8.31(c)** Select the Counterbore Side of the Plate

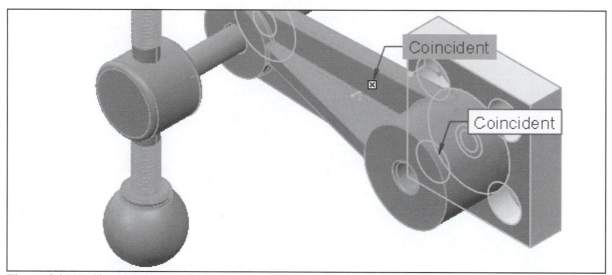

**Figure 8.31(d)** Clamp_Subassembly Reverses Position

**Figure 8.31(e)** Correct Clamp_Subassembly Orientation

Press: **MMB** spin the model > select the **CARRLANE-12-13350_STUD.PRT** in the Model Tree > **RMB** > **Edit Definition** > **Placement** tab > **Fix** > place the pointer on Fix, **RMB** > **Delete** [Fig. 8.32(a)] > **New Constraint** > pick the flat end surface of the stud > pick the bottom surface of the plate [Fig. 8.32(b)]

**Figure 8.32(a)** Delete Fix Constraint

**Figure 8.32(b)** Select the Bottom Surface of the Clamp_Plate

Click: in the Offset field > *type*: **-.25** > **Enter** [Fig. 8.32(c)] > **Enter** > **LMB** in the Graphics Window >
place the pointer on the short stud to highlight [Fig. 8.32(d)]

**Figure 8.32(c)** Offset Distance

**Figure 8.32(d)** Highlight the Short Stud

**Figure 8.32(e)** Standard Orientation

Click: **File > Manage File > Delete Old Versions > Enter >**  *off >* [ ] *off >* [ ] *off >* [ ] *off >* [ ]
> [ ] > [ Shading With Reflections ] (Fig. 8.33) > **File > Save As >** Type [ ▼ ] > **Zip File (*.zip) > OK >**
upload to your course interface or attach to an email and send to your instructor or yourself > **File > Close**

**Figure 8.33** Shading With Reflections *(the quality of your graphics card and graphics settings may prevent this display)*

A complete set of projects and illustrations are available at ***www.cad-resources.com.***

**Figure 9.1** Exploded Swing Clamp Assembly

## OBJECTIVES

- Create **Exploded Views**
- Create **Explode States**
- Utilize the **View Manager** to organize and control views states
- Create unique component **visibility settings (Style States)**
- **Move** and **Rotate** components in an assembly
- Add a **URL** to a **3D Note**
- Create **Perspective** views of the model

## REFERENCES AND RESOURCES

For **Resources** go to **www.cad-resources.com** > click on the Creo Parametric 2.0 Book cover

- Lesson 9 Lecture
- Lesson 9 3D models embedded in a PDF
- Project Lectures

## EXPLODED ASSEMBLIES

Pictorial illustrations, such as exploded views, are generated directly from the 3D model database (Fig. 9.1). The model can be displayed and oriented in any position. Each component in the assembly can have a different display type: wireframe, hidden line, no hidden, shading, and transparent. You can select and orient the component to provide the required view orientation to display the component from underneath or from any side or position. Perspective projections are made with selections from menus. The assembly can be spun around, reoriented, and even clipped to show the interior features. You have the choice of displaying all components and subassemblies or any combination of components in the design.

# Creating Exploded Views

Using the **Explode State** option in the **View Manager**, you can automatically create an exploded view of an assembly (Fig. 9.2). Exploding an assembly affects only the display of the assembly; it does not alter true design distances between components. Explode states are created to define the exploded positions of all components. For each explode state, you can toggle the explode status of components, change the explode locations of components, and create and modify explode offset lines to show how explode components align when they are in their exploded positions. The Explode State Explode Position functionality is similar to the Package/Move functionality.

You can define multiple explode states for each assembly and then explode the assembly using any of these explode states at any time. You can also set an explode state for each drawing view of an assembly. Creo Parametric 2.0 gives each component a default explode position determined by the placement constraints. By default, the reference component of the explode is the parent assembly (top-level assembly or subassembly).

To explode components, you use a drag-and-drop user interface similar to the Package/Move functionality. You select the motion reference and one or more components, and then drag the outlines to the desired positions. The component outlines drag along with the mouse cursor. You control the move options using a Preferences setting. Two types of explode instructions can be added to a set of components. The children components follow the parent component being exploded or they do not follow the parent component. Each explode instruction consists of a set of components, explode direction references, and dimensions that define the exploded position from the final (installed) position with respect to the explode direction references.

**Figure 9.2** Exploded Assemblies

When using the explode functionality, keep in mind the following:

- You can select individual parts or entire subassemblies from the Model Tree or main window.
- If you explode a subassembly, in the context of a higher-level assembly, you can specify the explode state to use for each subassembly.
- You do not lose component explode information when you turn the status off. Creo Parametric 2.0 retains the information so that the component has the same explode position if you turn the status back on.
- All assemblies have a default explode state called "Default Explode", which is the default explode state Creo Parametric 2.0 creates from the component placement instructions.
- Multiple occurrences of the same subassembly can have different explode characteristics at a higher-level assembly.

## Component Display

Style, also accessed through the View Manager, manages the display styles of an assembly. Simp Rep, Sections, Layers, Orient, Explode, and All are on separate tabs from this dialog [Figs. 9.3(a-h)]. Wireframe, hidden line, no hidden, shaded, or transparent display styles can be assigned to each component. The components will be displayed according to their assigned display styles in the current style state (that is, blanked, shaded, drawn in hidden line color, and so on). The current setting, in the View tab > Display Style, controls the display of unassigned components.

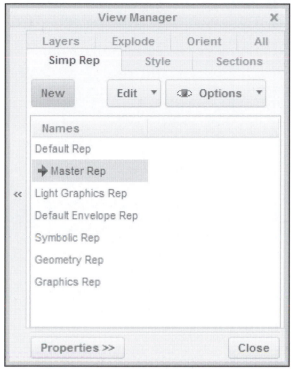

**Figure 9.3(a)** View Manager Dialog Box

**Figure 9.3(b)** View Manager- Simp Rep

**Figure 9.3(c)** View Manager- Style

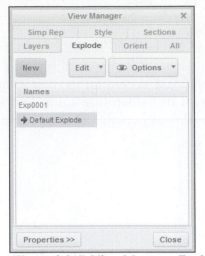

**Figure 9.3(d)** View Manager- Explode

**Figure 9.3(e)** View Manager- Layers

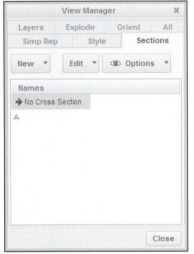

**Figure 9.3(f)** View Manager- Sections

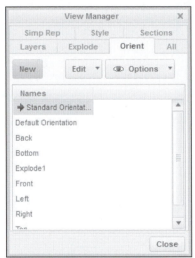

**Figure 9.3(g)** View Manager- Orient

**Figure 9.3(h)** View Manager- All

Components appear in the currently assigned style state [Fig. 9.4(a)]. The current setting is indicated in the display style column in the Model Tree window. Component display or style states can be modified without using the View Manager. You can select desired models from the graphics window, Model Tree, or search tool, and then use the View tab > Display Style commands to assign a display style (wireframe, hidden, no hidden, shaded and transparent) to the selected models. The Style representation is temporarily changed. These temporary changes can then be stored to a new style state, or updated to an existing style state. You can also define default style states. If the default style state is updated to reflect changes different from that of the master style state, then that default style state will be reflected each time the model is retrieved.

**Figure 9.4(a)** Style States

406

## Types of Representations

The main types of simplified representations: **default**, **master**, **light graphics** [Fig. 9.4(b)], **default envelope**, **symbolic**, **geometry** and **graphics**; designate which representation appears, using the commands in the View Manager dialog box. Graphics and geometry representations speed up the retrieval process of large assemblies. All simplified representations provide access to components in the assembly and are based upon the Master Representation. You cannot modify a feature in a graphics representation, but you can do so in a geometry representation.

**Figure 9.4(b)** Light Graphics Rep

Assembly features are displayed when you retrieve a model. Subtractive assembly features such as cuts and holes are represented in graphics and geometry representations, making it possible to use these simplified representations for performance improvement while still displaying on screen a completely accurate geometric model. You can access model information for graphics and geometry representations of part models from the Information menu and from the Model Tree. Because part graphics and geometry representations do not contain feature history of the part model, information for individual features of the part is not accessible from these representations.

- The *Master Representation* always reflects the fully detailed assembly, including all of its members. The Model Tree lists all components in the Master Representation of the assembly, and indicates whether they are included, excluded, or substituted.
- The *Graphics Representation* contains information for display only and allows you to browse through a large assembly quickly. You cannot modify or reference graphics representations. The type of graphic display available depends on the setting of the *save_model_display* configuration option, the last time the assembly was saved:
  - *wireframe* (default) The wireframe of the components appear.
  - *shading_low*, *shading_med*, *shading_high* A shaded version of the components appears. The different levels indicate the density of the triangles used for shading.
  - *shading_lod* The level of detail depends on the "Shaded model display settings" dialog box setting. Click: File > Options > Model Display before you open a model to access this dialog box.

While in a simplified representation, Creo Parametric 2.0 applies changes to an assembly, such as creation or assembly of new components, to the Master Representation. It reflects them in all of the simplified representations. It applies all suppressing and resuming of components to the Master Representation. However, it applies the actions of a simplified representation only to currently resumed members, which is, to members that are present in the BOM of the Master Representation.

CLAMP_ASSEMBLY.ASM
▶ Annotations
   ⧄ CL_ASM_RIGHT
   ⧄ CL_ASM_TOP
   ⧄ CL_ASM_FRONT
   ✳ CL_ASM_CSYS
▶ ⬜ CLAMP_PLATE.PRT
▼ ⬜ CLAMP_SUBASSEMBLY.ASM
   ▶ ⬜ Placement
   ⧄ ASM_RIGHT
   ⧄ ASM_TOP
   ⧄ ASM_FRONT
   ✳ SUB_ASM_CSYS
   ▶ ⬜ CLAMP_ARM_DESIGN<CLAMP
   ▶ ⬜ CLAMP_SWIVEL.PRT
   ▶ ⬜ CLAMP_FOOT.PRT
   ▶ ⬜ CARRLANE-12-13500_STUD.F
   ▶ ⬜ CLAMP_BALL.PRT
   ▶ ⬜ CLAMP_BALL.PRT
   ➡ Insert Here
▶ ⬜ CARRLANE-12-13350_STUD.PRT
▶ ⬜ CARRLANE_500_FN.PRT
➡ Insert Here

**Figure 9.5** Exploded Swing Clamp Assembly

## Exploded Swing Clamp Assembly

In this lesson, you will use the previously created subassembly and assembly (Fig. 9.5) to establish and save new views, exploded views, and views with component style states that differ from one another. The View Manager Dialog box will be employed to control and organize a variety of different states.

You will also be required to move and rotate components of the assembly before cosmetically displaying the assembly in an exploded state. The creation and assembly of new components will not be required. A 3D Note with a URL will be added to the model as the last information feature.

## URLs and Model Notes

Since the assembly you created is a standard clamp from CARR LANE Manufacturing Co.; you could go to CARR LANE's Website and see the assembly, order it (about $70.00 U.S.) or download a 3D IGES model. Here you will create and attach a 3D Note to the assembly, identifying the manufacturer and Website.

Launch **Creo Parametric 2.0** > **File** > **Manage Session** > **Select Working Directory** > select the directory where the **clamp_assembly.asm** was saved > **OK** > **File** > **Open** > **clamp_assembly.asm** > **Open** > **File** > **Options** > **Model Display** > **Trimetric** > **Isometric** > **OK** > **No** > 🏠 > 🌲 Tree Filters. > toggle all *on* > **OK** > **View** tab > [icons] *off* > [icons] *off* > [icon] *on* > [icon] *on* > **Layers** (from the Ribbon) > pick on 📚 Layers (in the Layer Tree) > **RMB** > **Hide** > **Layers** (from the Ribbon) > 🖼 > **Annotate** tab > [FLAT TO SCREEN / FRONT / TOP / RIGHT icons] (Fig. 9.6) > **RMB** (on the FLAT TO SCREEN icon) > **Set** > A≡ Notes (in the Annotations Group) > **New** from MDL NOTES in the Menu Manager > Name: *type* **CARR_LANE** > Text: *type text lines* (Fig. 9.7)

**Figure 9.6** FLAT TO SCREEN

**Figure 9.7** Note Dialog Box

409

Click: **Hyperlink** > *type* **www.carrlane.com** (Fig. 9.8) > **ScreenTip** > Screen tip text: *type* **CARR LANE Manufacturing** (Fig. 9.9) > **OK** > **OK** > **Place** >  pick near the assembly in the Graphics Window > **OK** > in the Graphics Window, press **RMB** > **Text Style** > Default > Height **4** > **Enter** (Fig. 9.10) > **Apply** > **OK** (Fig. 9.11) > **LMB** in the Graphics Window to deselect > **Ctrl+R** > **Ctrl+D** > **Ctrl+S** > **OK**

**Figure 9.8** Edit Hyperlink Dialog Box

**Figure 9.9** Set Hyperlink Screen Tip

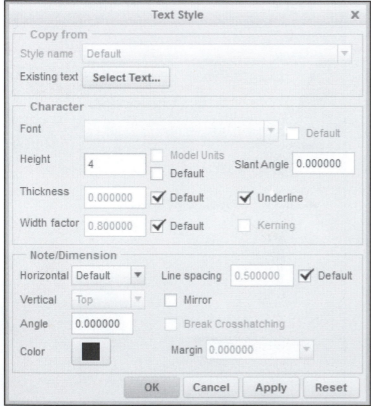

**Figure 9.10** Text Style Dialog Box

**Figure 9.11** Note Displayed in Model Tree and on Model

To open a Hyperlink defined in a model note and to launch the embedded Web browser and go to a World Wide Web URL (Universal Resource Locator) associated with a model note, use one of two methods.

Method 1: **Annotate** tab > pick on the note in the Model Tree or Graphics Window > press **RMB** [Fig.

9.12(a)] the shortcut menu displays > **Open URL** [Fig. 9.12(b)] the associated URL opens in the Creo Parametric 2.0 embedded browser *(if you get an error: delete +0 or +1 from the end of the URL (**http://www.carrlane.com/?CLAMP_ASSEMBLY.ASM** > Enter)* > navigate the site to locate the Swing Clamp [Figs. 9.12(c-g)] > **Dimensions** tab > use the scroll bars to the see the drawing [Figs. 9.12(h-i)])
*(Note that to pick the 3D Note in the Graphics Window; you may have to change the Selection Filter to Annotation)*

Method 2: Move the cursor over the hyperlinked note > press **Ctrl** and the cursor changes to a hand icon > click **LMB** and the associated URL opens in the Creo Parametric 2.0 embedded browser *(if you get an error: delete +0 or +1 from the end of the URL (**http://www.carrlane.com/?CLAMP_ASSEMBLY.ASM** > Enter)* > navigate the site to locate the Swing Clamp

**Figure 9.12(a)** Open URL

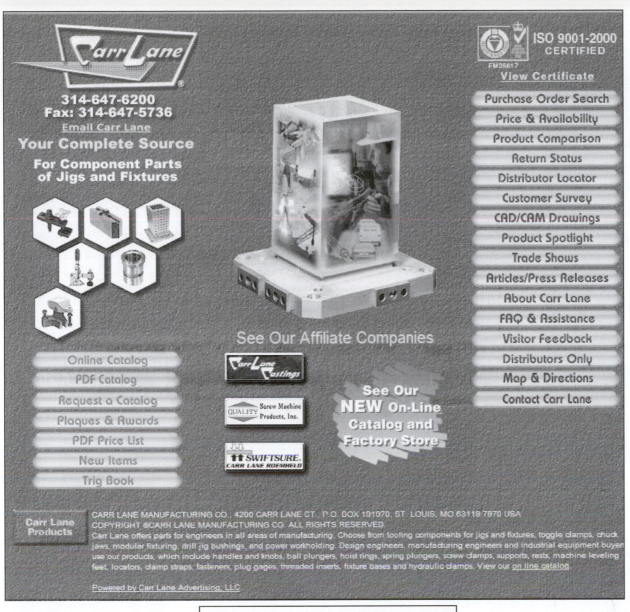

**Figure 9.12(b)** Carr Lane Web Site  http://www.carrlane.com/?CLAMP_ASSEMBLY.ASM

**Figures 9.12(c-g)** Online Catalog > Clamps and Accessories > Swing Clamps > Swing Clamp Assemblies > Flange Mounted Ball Handle

413

Features                              Dimensions

# SWING CLAMP ASSEMBLIES

**FEATURES:** Long-reach swing clamp that can be hand tightened for medium clamping force. This flange-mounted version is ideal for mounting on flat surfaces. The stop pin furnished can be installed on either side for clockwise or counterclockwise swing.

**SIZES:** Available in five sizes — see Dimensions page for part numbers:
- 1-1/2" Reach
- 2" Reach
- 2-1/2" Reach
- 3" Reach
- 4" Reach

**SWIVEL FOOT:** Swing Clamp Assemblies use our patented left-hand-thread swivel-foot system. The ball end of the body screws into the foot for easy installation and removal. Once the ball threads pass beyond the foot threads, the foot can rotate and pivot freely (20° total angle). Foot threads are left-hand to positively prevent accidental black out during use. Delrin® feet are optional. To order with a Delrin® foot, add "N" to the part number, e.g. CL-1-SWA-3N. Note: Delrin® feet are not threaded — they snap on and off. For other foot types, see Feet for Swivel Screws (order separately).

**MATERIAL:** Swing arm > Nodular iron per ASTM A536 grade 65-45-12, zinc plated clear chromate. Post and flange base > Mild steel, black oxide finish. Other components > See catalog information for each specific item.

Flange Mounted
Ball Handle

en_2063

**MANUFACTURING CO.**
4200 Carr Lane Ct., P.O. Box 191970
St. Louis, Missouri 63119-7970 USA
Phone: 314-647-6200, FAX: 314-647-5736
Web Site: www.carrlane.com

## USA Catalog
General Information
Alignment Pins
Hoist Rings
Handles/Knobs/Screw Clamps
Spring-Loaded Devices
Supports, Rests, and Feet
Locators
Clamps and Accessories
Plug Gages
Threaded Inserts
Jig and Fixture Bases
Chuck Jaws
Modular Fixturing
Toggle Clamps
Drill-Jig Bushings
Power Workholding

SEARCH

New Items
Find a Carr Lane Distributor
View My Shopping Cart
CAD Drawings Search
Part Number Conversion
Trig Book
Carr Lane Home Page

**Figure 9.12(h)** Ball Handle Flange Mounted Swing Clamp

USA Catalog

General Information

Alignment Pins

Hoist Rings

Handles/Knobs/Screw Clamps

Spring-Loaded Devices

Supports, Rests, and Feet

Locators

Clamps and Accessories

Plug Gages

Threaded Inserts

Jig and Fixture Bases

Chuck Jaws

Modular Fixturing

Toggle Clamps

Drill-Jig Bushings

Power Workholding

SEARCH

New Items

Find a Carr Lane Distributor

View My Shopping Cart

CAD Drawings Search

Part Number Conversion

Trig Book

Carr Lane Home Page

Features                          Dimensions

# SWING CLAMP ASSEMBLIES

Add "N" to part number for Delrin foot, e.g. CL-1-SWA-4N.

FLANGE MOUNTED, BALL HANDLE

| PART NO. | CL-1-SWA-4 | CL-4-SWA-4 | CL-5-SWA-4 | CL-2-SWA-4 | CL-3-SWA-4 |
|---|---|---|---|---|---|
| D THREAD SIZE | 5/16-18 | 1/2-13 | 1/2-13 | 5/8-11 | 5/8-11 |
| A | 1-1/2 | 2 | 2-1/2 | 3 | 4 |
| B | 1 | 1-3/8 | 1-3/8 | 1-1/2 | 1-1/2 |
| C | 3/4 | 1-1/8 | 1-1/8 | 1-1/4 | 1-1/4 |
| E | .88 | .96 | .96 | 1.65 | 1.65 |
| F | 3/4 | 1-1/8 | 1-1/8 | 1-5/16 | 1-5/16 |
| G | 2.39 | 3.53 | 3.53 | 4.50 | 4.50 |
| H DIA | 1/2 | 3/4 | 3/4 | 15/16 | 15/16 |
| J | 2-1/4 | 2-5/8 | 2-5/8 | 3 | 3 |
| K | 5/32 | 3/16 | 3/16 | 7/32 | 7/32 |
| L | .33 | .44 | .44 | .58 | .58 |
| M | 1/2 | 1/2 | 11/16 | 11/16 | 1 |
| N | .25 | .31 | .31 | .53 | .53 |

**Figure 9.12(i)** Select Dimensions Tab

Click: [×] to close the Browser > [icon] **Annotation Display** *off* > **Ctrl+D** > **Ctrl+S** > **Enter** > in the Graphics Window, **LMB** to deselect

# Views: Perspective, Saved, and Exploded

You will now create a variety of cosmetically altered view states. Cosmetic changes to the assembly do not affect the model itself, only the way it is displayed on the screen. One type of view that can be created is the perspective view.

Perspective creates a single-vanishing-point perspective view of a shaded or wireframe model. These views allow you to observe an object as the view location follows a curve, axis, cable, or edge through or around an object. To add perspective to a model view, you select a viewing path and then control the viewing position along the path in either direction. You can also rotate the perspective view in any direction, zoom the view in or out, and change the view angle at any point along the path.

Click: 🔍 **Command Search** > *type* **pers** > [Perspective View] > [Perspective ▼] > [Perspective Settings] [Fig. 9.13(a)] > [Perspective View Settings ▼] [Fig. 9.13(b)] > examine the options > **LMB** in the Graphics Window > [35 mm - Standard ▼] > examine the options > **LMB** in the Graphics Window > adjust the settings as desired > **OK** > ⓘ > in the search box, *type* **view perspective** > **Enter** > [About Adding **Perspective** to a **View**] > read the Help file > [×] > **Ctrl+D** > **Ctrl+S** > **Enter**

**Figure 9.13(a)** Perspective Dialog Box

**Figure 9.13(b)** Perspective View

416

## Saved Views

Create a saved view and explode state to be used later on an assembly drawing. When using the View Manager to set display and explode states, it is a good idea to create one or more saved views to be used later for exploding. The default trimetric and isometric views do not adequately represent the assembly in its functional position.

Using the **MMB**, rotate the model from its default position [Fig. 9.14(a)] to one where the Clamp_Swivel is approximately vertical [Fig. 9.14(b)] >  from the Graphics Toolbar > [Reorient] [Fig. 9.14(c)] > **Saved Views** [Fig. 9.14(d)] > Name: *type* **EXPLODE1** > **Save** > **OK** [Fig. 9.14(e)]

**Figure 9.14(a)** Isometric View

**Figure 9.14(b)** Reoriented View

**Figure 9.14(c)** Orientation Dialog Box

**Figure 9.14(d)** Saved Views

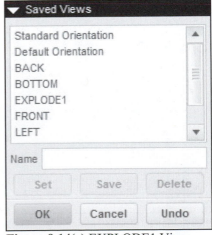

**Figure 9.14(e)** EXPLODE1 View

417

## Default Exploded Views

When you create an *exploded view*, Creo Parametric 2.0 moves apart the components of an assembly to a set default distance. The default position is seldom the most desirable.

Click: **Model** tab > Exploded View [Figs. 9.15(a-b)] > 🔍 > Exploded View > 🔍

**Figure 9.15(a)** Exploded View

**Figure 9.15(b)** Default Exploded View

Click: **View** tab > [Exploded View] > [🔍] > [Exploded View] > [🔍] > place the pointer inside the Graphics Toolbar > **RMB** > **Location** > **Show in Status Bar** [Fig. 9.16(a)] > press **LMB** on the bottom edge of the Graphics Window ([be displaye]) > move the pointer to change the height of the message area [Fig. 9.16(b)]

Figure 9.16(a) Move the Location of the Graphics Toolbar to the Status Bar

- Datum planes will not be displayed.
- Axes will be displayed.
- Axes will not be displayed.
⚠️ Animation frames have been skipped.
- Assembly components have been successfully exploded.
⚠️ Animation frames have been skipped.
- Assembly components have been successfully unexploded.
⚠️ Animation frames have been skipped.
- Assembly components have been successfully exploded.
⚠️ Animation frames have been skipped.

Figure 9.16(b) Resized Message Log and Status Bar with Integrated Graphics Toolbar *(yours may appear differently)*

Click: **Model** tab >  **View Manager** from the Status Bar > **Simp Rep** tab > **Master Rep** > **Options** > **Activate** > **Orient** tab > **Explode1(+)** > **RMB** > **Save** [Fig. 9.17(a)] > **OK** [Fig. 9.17(b)] > **Explode** tab > **New** > **Enter** to accept the default name: Exp0001 [Fig. 9.17(c)] > **Properties**>> (bottom left of the View Manager dialog box) > **Edit position** [Fig. 9.17(d)] > **References** tab [Fig. 9.17(e)]

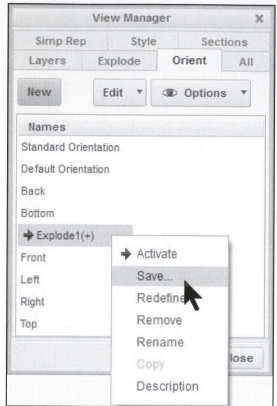

**Figure 9.17(a)** View Manager Orient Tab

**Figure 9.17(b)** Save Display Elements

**Figure 9.17(c)** Exp0001    **Figure 9.17(d)** Edit Position    **Figure 9.17(e)** Explode Position Dashboard

Click: **Options** tab > 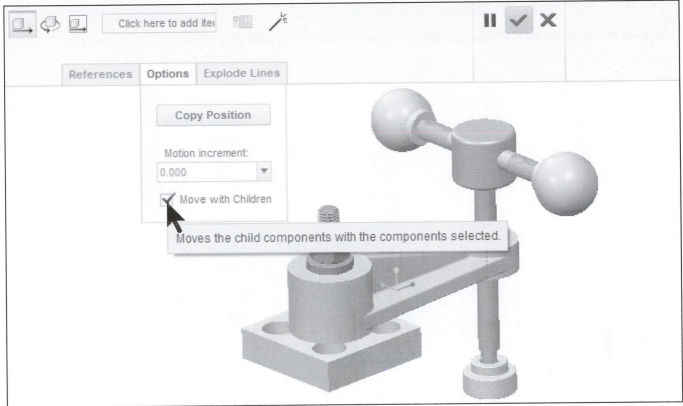 ✔ Move with Children [Fig. 9.18(a)] > **References** tab > Components to Move: select the top of the **CARRLANE-12-13350_STUD** [Fig. 9.18(b)]

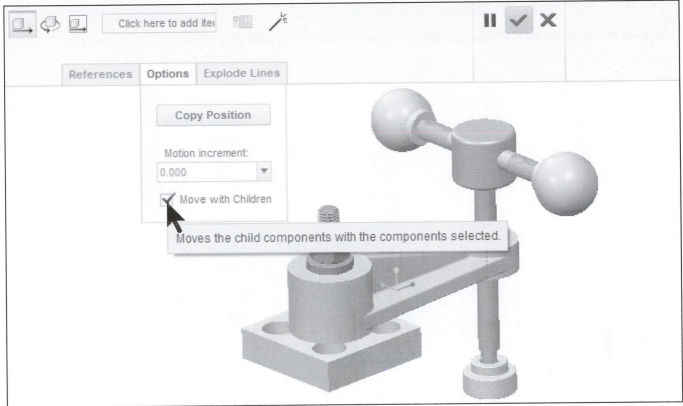

**Figure 9.18(a)** Options: Move with Children

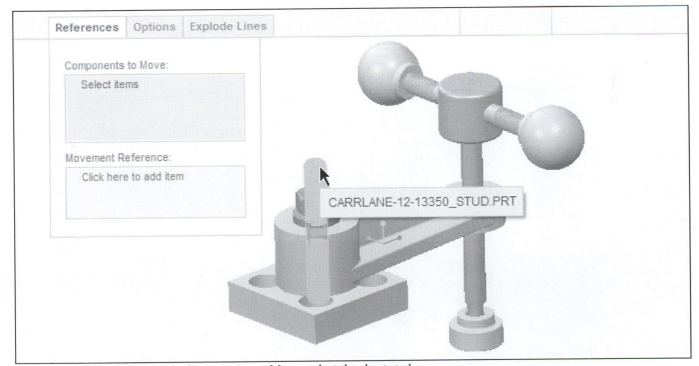

**Figure 9.18(b)** References tab, Components to Move: select the short stud

Using the **LMB**, select and hold the **X axis** of the temporary coordinate system [Fig. 9.18(c)] > move the pointer to drag the components to a new position > release the **LMB** [Fig. 9.18(d)]

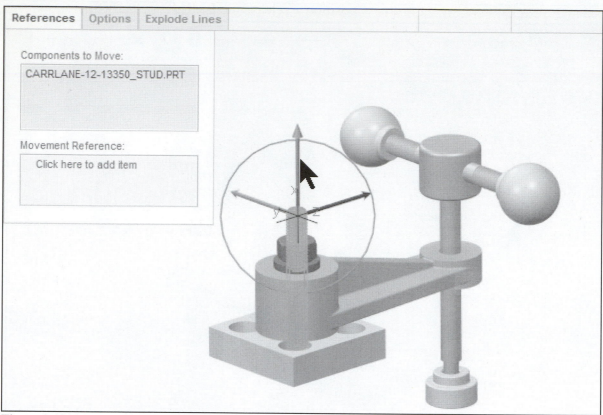

**Figure 9.18(c)** Press and hold the LMB on the X Axis *(your view may appear differently)*

**Figure 9.18(d)** Move the components to a new position *(your view may appear differently)*

Click: **Options** tab > ☐ Move with Children > **References** tab > Components to Move: select the bottom corner of the **CLAMP_PLATE** [Fig. 9.18(e)] > press **LMB** on the vertical axis of the temporary coordinate system > drag the component to a new position > release the **LMB** [Figs. 9.18(f-g)]

**Figure 9.18(e)** Select CLAMP_PLATE

**Figure 9.18(f)** Drag Component by Coordinate System Axis

**Figure 9.18(g)** Moving CLAMP_PLATE > at the Desired Position, Release the LMB

Select **CARRLANE_500_FN** [Fig. 9.18(h)] > press **LMB** on the **X Axis** of the temporary coordinate system > drag to a new position [Fig. 9.18(i)] > release the **LMB** [Fig. 9.18(j)]

**Figure 9.18(h)** Select the Flange Nut

**Figure 9.18(i)** Drag the Flange Nut to a New Position

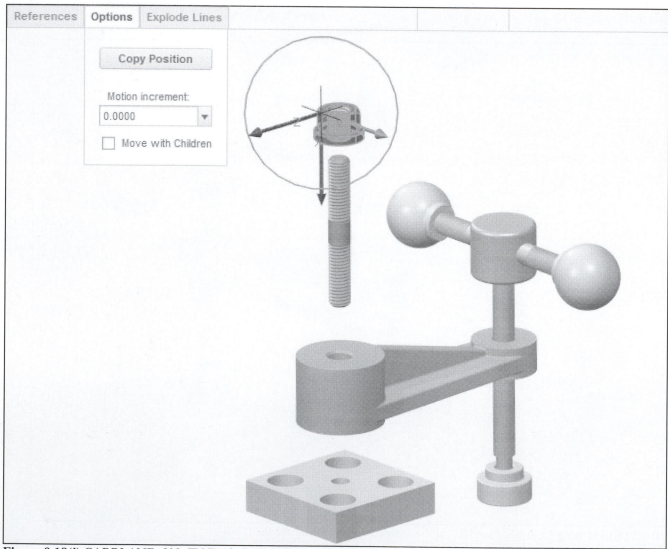

**Figure 9.18(j)** CARRLANE_500_FN Explode Position

Click: **Options** tab > ☑ Move with Children > **References** tab > Components to Move: **select CLAMP_SWIVEL** [Fig. 9.18(k)] > press **LMB** on the vertical axis > drag the component to a new location [Fig. 9.18(l)] > release the **LMB** > pan and zoom as required

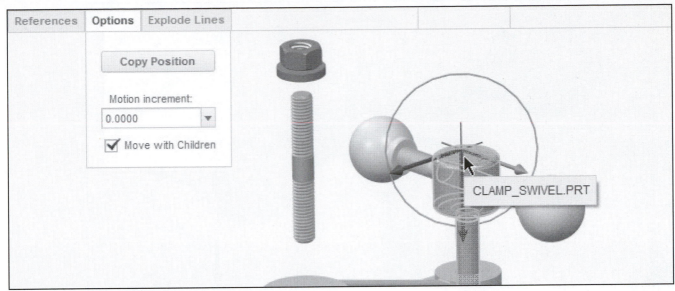

**Figure 9.18(k)** Select the CLAMP_SWIVEL

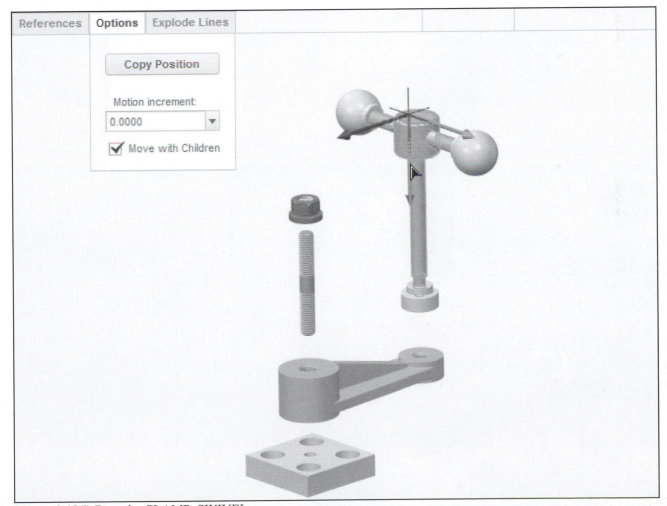

**Figure 9.18(l)** Drag the CLAMP_SWIVEL

Click: **Options** tab > ☐ Move with Children > **References** tab > Components to Move: select the **CLAMP_BALL** [Fig. 9.18(m)] > press **LMB** on the axis > move the component [Fig. 9.18(n)] > release the **LMB** > repeat the process and move the second clamp ball and the long stud [Fig. 9.18(o)]

**Figure 9.18(m)** Select and Move the CLAMP_BALL

**Figure 9.18(n)** Moved CLAMP_BALL

**Figure 9.18(o)** Select and Move the Second CLAMP_BALL and the CARRLANE-12-13500_STUD

Select the **CLAMP_FOOT** > press **LMB** on the axis of the temporary coordinate system > move the component [Fig. 9.18 (p)] > release the **LMB** > ☑ > 🔍 > in the Graphics Window, **LMB** to deselect [Fig. 9.18 (q)]

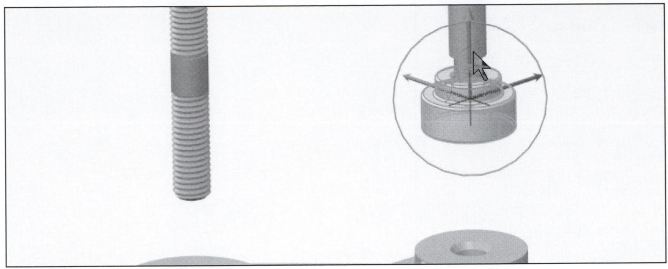

**Figure 9.18(p)** Move the CLAMP_FOOT

**Figure 9.18(q)** View Manager Explode Tab

Click: **<<List > Exp0001(+) > RMB > Save** [Fig. 9.18(r)] **> OK** [Fig. 9.18(s)] **> Orient** tab > **Explode1(+) > Edit > Save > OK > Close**

**Figure 9.18(r)** Save the Explode State

**Figure 9.18(s)** Save Display Elements Dialog Box

## View Style

The components of an assembly (whether exploded or not) can be displayed individually with **Wireframe**, **Hidden Line**, **No Hidden**, **Shading**, or **Transparent**. Style is used to manage the display styles of an assembly's components. The components will be displayed according to their assigned display style in the current style state.

Click: [icon] **View Manager** Open the View Manager > **Style** tab > **New** > **Enter** [Fig. 9.19(a)] to accept the default name: Style0001 > **Show** tab [Fig. 9.19(b)] > [◉ Wireframe] > select the **CLAMP_SWIVEL** > select the **CARRLANE-12-13350_STUD** > [👓] **Update**

**Figure 9.19(a)** Style State Style0001

**Figure 9.19(b)** Edit Dialog Box, Show Tab, Wireframe

429

Click:  Hidden Line > press and hold down the **Ctrl** key > select the **CLAMP_PLATE** and the **CLAMP_FOOT** > release the **Ctrl** key > ✓ ✗ ᨑ **Update** [Fig. 9.19(c)]

**Figure 9.19(c)** Edit Dialog Box, Show Tab, Hidden Line

430

Click:  No Hidden > select the two **CLAMP_BALL** components > 👓 **Update Model View According To Specified Changes** > ⦿ Shading > select **CARRLANE_500_FN** and **CARRLANE-12-13500_STUD** > 👓 **Update** [Fig. 9.19(d)]

**Figure 9.19(d)** Edit Dialog Box, Show Tab, Shading

431

Click:  ⦿ Transparent > select **CLAMP_ARM_DESIGN** > 👓 **Update Model View According To Specified Changes** [Fig. 9.19(e)] > ✓ ✗ 👓

**Figure 9.19(e)** Completed Style State

Click: **Orient** tab > {*if* (+) next to **Explode1** > **Edit** > **Save** > **OK** [Fig. 9.19(f)]} > **All** tab > ☑ Display combined views > **New** > **Enter** to accept the default name: Comb0001 [Fig. 9.19(g)] > **Reference**

**New Presentation State** ✕

Would you like to reference existing display states or create new copies?

[ Reference Originals ] [ Create Copies ]

**Originals** > **Edit** > **Preferences** [Figs. 9.19(h-i)] > **OK** > **Close**

**Figure 9.19(f)** Orient Tab

**Figure 9.19(g)** All Tab

**Figure 9.19(h)** Preferences

**Figure 9.19(i)** All Edit Preferences

433

## Model Tree

You can display component style states, and explode states in the Model Tree.

Click:  from the Navigator > **Tree Columns** > **Info** > **Display Styles** [Fig. 9.20(a)] > **STYLE0001** > `>>` [Fig. 9.20(b)] > **Apply** > **OK** > adjust the column format [Fig. 9.20(c)]

**Figure 9.20(a)** Model Tree Columns

**Figure 9.20(b)** Adding Display Style STYLE0001 to the Model Tree Columns

**Model Tree**

| | Display St... |
|---|---|
| CLAMP_ASSEMBLY.ASN | |
| ▶ Annotations | |
| ▱ CL_ASM_RIGHT | |
| ▱ CL_ASM_TOP | |
| ▱ CL_ASM_FRONT | |
| ✳ CL_ASM_CSYS | |
| ▶ CLAMP_PLATE.PRT | Hidden line |
| ▼ CLAMP_SUBASSEN | |
|   ▶ Placement | |
|   ▱ ASM_RIGHT | |
|   ▱ ASM_TOP | |
|   ▱ ASM_FRONT | |
|   ✳ SUB_ASM_CSY | |
|   ▶ CLAMP_ARM_D | Transparent |
|   ▶ CLAMP_SWIVEL | Wireframe |
|   ▶ CLAMP_FOOT.P | Hidden line |
|   ▶ CARRLANE-12- | |
|   ▶ CLAMP_BALL.P | No Hidden |
|   ▶ CLAMP_BALL.P | No Hidden |
|   ➔ Insert Here | |
| ▶ CARRLANE-12-133 | Wireframe |
| ▶ CARRLANE_500_F | |
| ➔ Insert Here | |

Explode State: EXP0002
Style State: STYLE0001

**Figure 9.20(c)** Adjusted Model Tree

Click:  *off* > **View** tab > [Default All] (at the bottom of the Graphics Window) > [Comb0001] [Fig. 9.20(d)] [Explode State:EXP0001 / Style State:STYLE0001] > [Model Display ▼] **Model Display Group** > [🗔 Temporary Shade] > [Style State:DEFAULT STYLE / On-Demand Simp Rep:DEFAULT REP] > **Ctrl+S** > **Enter** > **File** > [Default All] [Fig. 9.20(e)]

**Manage File** > **Delete Old Versions** > **Enter** > **File** > **Save As** > Type [▾] > **Zip File (\*.zip)** > **OK** > **Cancel** (in the Duplicate File Name Found dialog box) > New Name: *type* **clamp_assembly_exp.asm.zip** > **OK** > **Yes** (in the Warning: dialog box) > upload to your course interface or attach to an email and send to your instructor or yourself > **File** > **Close** > **File** > **Exit** > **Yes**

Explode State:EXP0001
Style State:STYLE0001

**Figure 9.20(d)** Comb0001

Style State:DEFAULT STYLE
On-Demand Simp Rep:DEFAULT REP

**Figure 9.20(e)** Default All

You have completed the Assembly Mode lessons. The assembly and view states created in Lessons 8 and 9 will be used in Lesson 12 to create assembly drawings. The component parts modeled in Lessons 6 and 7 will be used in Lesson 10 to create detail drawings.

Download additional projects from ***www.cad-resources.com.***

436

**Figure 10.1** Swing Clamp Assembly Component Drawings

## OBJECTIVES

- Create part drawings
- Display **standard views** using the template
- **Add, Move, Erase**, and **Delete** views
- Retrieve a standard **format**
- **Display** Gtol datums, centerlines, and dimensions
- **Cleanup** dimension positions
- Specify and retrieve standard **Format** paper size and units
- Change the **Scale** of a view

## REFERENCES AND RESOURCES

For **Resources** go to **www.cad-resources.com** > click on the Creo Parametric 2.0 Book cover

- Lesson 10 Lecture
- Lesson 10 3D models embedded in a PDF
- Book Projects PDF

## Introduction to Drawings

This lesson will introduce the basic concepts and procedure for creating detail drawings (Fig. 10.1). Using parts modeled for the Swing Clamp Assembly, you will quickly create drawings, change formats, delete and add views, and display dimensions and centerlines. Since this lesson is meant to introduce drawing mode concepts and procedures, a minimum of cleanup, reformatting and repositioning of dimensions and centerlines will be required.

# FORMATS, TITLE BLOCKS, AND VIEWS

**Formats** are user-defined drawing sheet layouts. A drawing can be created with an empty format and have a standard size format added as needed. Formats can be added to any number of drawings. The format can also be modified or replaced in a Creo Parametric 2.0 drawing at any time.

**Title Blocks** are standard or sketched line entities that can contain parameters (object name, tolerances, scale, and so on) that will show when the format is added to the drawing.

**Views** created by Creo Parametric 2.0 are identical to views constructed manually by a designer on paper. The same rules of projection apply; the only difference is that you choose commands in Creo Parametric 2.0 to create the views as needed.

## Formats

Formats consist of draft entities, not model entities. There are two types of formats: standard and sketched. You can select from a list of **Standard Formats** (**A-F** and **A0-A4**) or create a new size by entering values for length and width.

**Sketched Formats** created in Sketcher mode (Fig. 10.2) may be parametrically modified, enabling you to create nonstandard-size formats or families of formats.

Formats can be altered to include note text, symbols, tables, and drafting geometry, including drafting cross sections and filled areas.

**Figure 10.2** Sketched Format

With Creo Parametric 2.0, you can do the following in Format mode:

- Create draft geometry and notes (Fig. 10.3)
- Move, mirror, copy, group, translate, and intersect geometry
- Use and modify the draft grid
- Enter user attributes
- Create drawing tables
- Use interface tools to create plot, DXF, SET, and IGES files
- Import IGES, DXF, and SET files into the format
- Create user-defined line styles
- Create, use, and modify symbols
- Include drafting cross sections in a format

Whether you use a standard format or a sketched format, the format is added to a drawing that is created for a set of specified views of a parametric 3D model.

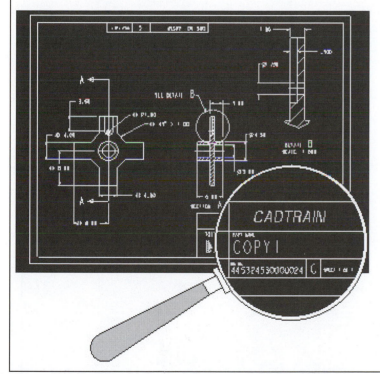

**When you place a format on your drawing, the system automatically writes the appropriate notes based on information in the model you use.**

**Figure 10.3** Format with Parametric Notes, Added to a Drawing

## Specifying the Format Size when Creating a New Drawing

If you want to use an existing template, select a listed template [Fig. 10.4(a)], or select a template from the appropriate directory. If you want to use the existing format [Fig. 10.4(b)], using Browse will open Creo Parametric 2.0's System Formats folder [Fig. 10.4(c)]. Select from the list of standard formats or navigate to a directory with user or company created formats.

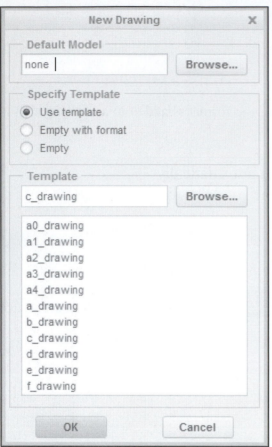

**Figure 10.4(a)** New Drawing Dialog Box (Use Template)

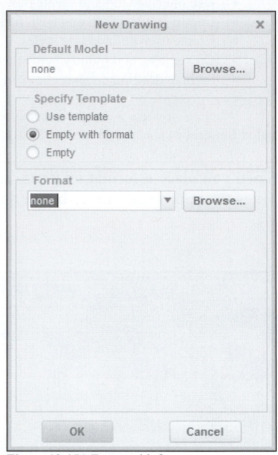

**Figure 10.4(b)** Empty with format

**Figure 10.4(c)** System Formats

If you want to create your own variable size format [Figs. 10.4(d-e)], enter values for width and length. The main grid spacing and format text units depend on the units selected for a variable size format. The New Drawing dialog box also provides options for the orientation of the format sheet:

- **Portrait** Uses the larger of the dimensions of the sheet size for the format's height; uses the smaller for the format's width
- **Landscape** Uses the larger of the dimensions of the sheet size for the format's width; uses the smaller for the format's height [Fig. 10.4(d)]
- **Variable** Select the unit type, Inches or Millimeters, and then enter specific values for the Width and Height of the format [Fig. 10.4(e)]

| | | | | | |
|---|---|---|---|---|---|
| **A0** | 841 X 1189 | mm | **A** | 8.5 X 11 | in. |
| **A1** | 594 X 841 | mm | **B** | 11 X 17 | in. |
| **A2** | 420 X 594 | mm | **C** | 17 X 22 | in. |
| **A3** | 297 X 420 | mm | **D** | 22 X 34 | in. |
| **A4** | 210 X 297 | mm | **E** | 34 X 44 | in. |
| | | | **F** | 28 X 40 | in. |
| | | | | | |

**Figure 10.4(d)** New Drawing Dialog Box (Empty- Landscape)

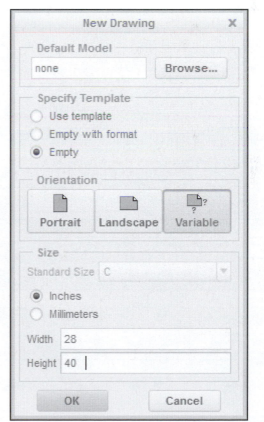

**Figure 10.4(e)** New Drawing (Empty- Variable)

441

# Drawing User Interface

The Drawing user interface helps you complete your detailing tasks quickly and efficiently. The **Ribbon** optimizes the user interface according to tasks. It consists of tabs with groups of frequently-used commands organized in a logical sequence. The group overflow area contains additional commands that are not frequently used. When you click a tab, the groups on the Ribbon, the drawing object filter, the shortcut menus, and the Drawing Tree update to enable operations relevant to the task. You can minimize or maximize the Ribbon display by double-clicking on a tab or by right-clicking on the Ribbon, and clicking Minimize the Ribbon on the shortcut menu. If you have minimized the Ribbon, click on a tab to temporarily maximize its display. In this case when you try to execute a command on the Ribbon or press the ESC key, the Ribbon changes back to its minimized state.

The **Navigation area** contains the Drawing Tree and the Model Tree. The Drawing Tree updates dynamically to reflect drawing objects relevant to the currently active tab. For example, when the Annotate tab is active, the annotations on the current sheet are listed in the Drawing Tree. In the navigation area, you can toggle between the Drawing Tree and the Layer Tree. Similarly you can toggle between the Model Tree and the Layer Tree. However, you cannot display the Layer Tree simultaneously in both the Drawing Tree area as well as the Model Tree area.

The **Drawing sheet** tabs are located below the graphics window. You can create multiple drawing sheets and move items from one sheet to another. You can also move or copy, add, rename, or delete a sheet.

## Drawing Templates

Drawing templates may be referenced when creating a new drawing. Templates can automatically create the views, set the desired view display, create snap lines, and show model dimensions. Drawing templates contain three basic types of information for creating new drawings. The first type is basic information that makes up a drawing but is not dependent on the drawing model, such as notes, symbols, and so forth. This information is copied from the template into the new drawing. The second type is instructions used to configure drawing views and the actions that are performed on their views. The instructions are used to build a new drawing object. The third type is a parametric note. Parametric notes are notes that update to new drawing model parameters and dimension values. The notes are re-parsed or updated when the template is instantiated. Use the templates to:

- Define the layout of views
- Set view display
- Place notes
- Place symbols
- Define tables
- Create snap lines
- Show dimensions

### Template View

You can also create customized drawing templates for the different types of drawings that you construct. Creating a template allows you to establish portions of drawings automatically, using the customizable template. The Template View Instructions dialog box (Fig. 10.5) is accessed through Tools tab > Template > Layout tab > Template View, when in the Drawing mode.

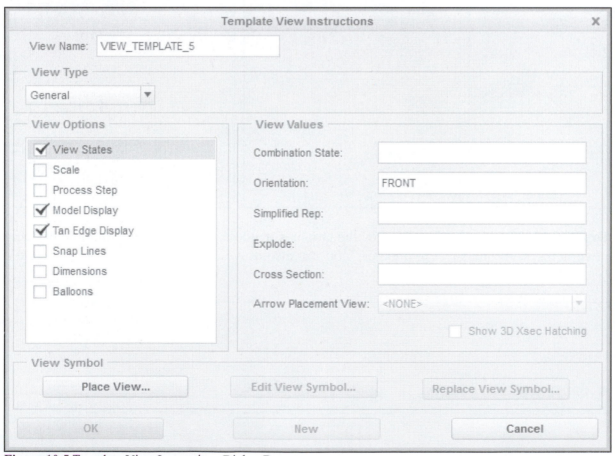

**Figure 10.5** Template View Instructions Dialog Box

You can use the following options in the Template View Instructions dialog box to customize your drawing templates:

- **View Name** Set the name of the drawing view that will be used as the view symbol label
- **View Orientation** Create a General view or a Projection view
- **Model "Saved View" Name** Orient the view based on a name view in the model
- **Place View** Places the view after you have set the appropriate options and values
- **Edit View Symbol** Allows you to edit the view symbol using the Symbol Instance dialog box
- **Replace View Symbol** Allow you to replace the view symbol using the Symbol Instance dialog box

| View Options | View Values |
|---|---|
| **View States**—Specify the type of view.<br><br>By default, the **View States** check box is selected and the view state is displayed when the **Template View Instructions** dialog box opens. The value of **Orientation** is FRONT, by default. If you select **Combination State**, the **Orientation**, **Simplified Rep**, **Explode** and **Cross Section** boxes display the text Defined by Combination State. The **Arrow Placement View** and **Show X-Hatching** check boxes also become available for selection after you specify the Combination State. | Combination State, Orientation, Simplified Rep, Explode, Cross Section, Arrow Placement View |
| **Scale**—Type a new value for the scale or use the default value. | View Scale |
| **Process Step**—Set the process step for the view.<br><br>You can specify the step number and set the view of the tool by checking the **Tool View** check box under **Process Step**. | Step Number, Tool View |
| **Model Display**—Set the view display for the drawing view. | Follow Environment, Wireframe, Hidden Line, No Hidden, Shading |
| **Tan Edge Display**—Set the tangent edge display. | Tan Solid, No Disp Tan, Tan Ctrln, Tan Phantom, Tan Dimmed, Tan Default |
| **Snap Lines**—Set the number, spacing, and offset of the snap lines. | Number, Incremental Spacing, Initial Offset |
| **Dimensions**—Show dimensions on the view. | Create Snap Lines, Incremental Spacing, Initial Offset |
| **Balloons**—Show balloons on the view. | |

## Views

A wide variety of views (Fig. 10.6) can be derived from the parametric model. Among the most common are projection views. Creo Parametric 2.0 creates projection views by looking to the left of, to the right of, above, and below the picked view location to determine the orientation of a projection view. When conflicting view orientations are found, you are prompted to select the view that will be the parent view. A view will then be constructed from the selected view.

At the time when they are created, projection, auxiliary, detailed, and revolved views have the same representation and explosion offsets, if any, as their parent views. From that time onward, each view can be simplified, be restored, and have its explosion distance modified without affecting the parent view. The only exception to this is detailed views, which will always be displayed with the same explosion distances and geometry as their parent views.

Once a model has been added to the drawing, you can place views of the model on a sheet. When a view is placed, you can determine how much of the model to show in a view, whether the view is of a single surface or shows cross sections, and how the view is scaled. You can then show the associative dimensions passed from the 3D model, or add reference dimensions as necessary. Basic view types used by Creo Parametric 2.0 include general, projection, auxiliary, and detailed:

- **General** Creates a view with no particular orientation or relationship to other views in the drawing. The model must first be oriented to the desired view orientation established by you.
- **Projection** Creates a view that is developed from another view by projecting the geometry along a horizontal or vertical direction of viewing (orthographic projection). The projection type is specified by you in the drawing setup file and can be based on third-angle (default) or first-angle rules.
- **Auxiliary** Creates a view that is developed from another view by projecting the geometry at right angles to a selected surface or along an axis. The surface selected from the parent view must be perpendicular to the plane of the screen.
- **Detailed** Details a portion of the model appearing in another view. Its orientation is the same as that of the view it is created from, but its scale may be different so that the portion of the model being detailed can be better visualized.
- **Revolved** A revolved view is a cross section of an existing view, revolved 90 degrees around a cutting plane projection. You can use a cross section created in the 3D model as the cutting plane, or you can create one on the fly while placing the view. The revolved view differs from a cross-sectional view in that it includes a line noting the axis of revolution for the view.

**Figure 10.6** Views

The view options that determine how much of the model is visible in the view are:

- **Full View** Shows the model in its entirety.
- **Half View** Removes a portion of the model from the view on one side of a cutting plane.
- **Broken View** Removes a portion of the model from between two selected points and closes the remaining two portions together within a specified distance.
- **Partial View** Displays a portion of the model in a view within a closed boundary. The geometry appearing within the boundary is displayed; the geometry outside of it is removed.

The options that determine whether the view is of a single surface or has a cross section are:

- **Section** Displays an existing cross section of the view if the view orientation is such that the cross-sectional plane is parallel to the screen (Fig. 10.7).
- **No Xsec** Indicates that no cross section is to be displayed.
- **Of Surface** Displays a selected surface of a model in the view. The single-surface view can be of any view type except detailed.

The options that determine whether the view is scaled are:

- **Scale** Allows you to create a view with an individual scale shown under the view. When a view is being created, Creo Parametric 2.0 will prompt you for the scale value. This value can be modified later. General and detailed views can be scaled.
- **No Scale** A view will be scaled automatically using a pre-defined scale value.
- **Perspective** Creates a perspective general view.

**Figure 10.7** Drawing with Section Views

447

**Figure 10.8** Clamp Arm Drawing

## Creating a Drawing of the Clamp Arm

Before starting the drawing you will need to set your working directory to the folder that contains the Swing Clamp Assembly (and therefore all the components of the assembly). The standard three views of the Clamp_Arm (Fig. 10.8) will display automatically on the drawing, since you will use the default drawing template for this project.

Launch **Creo Parametric 2.0** > **Select Working Directory** > select the directory where the **clamp_arm.prt** was saved > **OK** >  > Drawing > Name **clamp_arm** > ☑ Use default template [Fig. 10.9(a)] > **OK** [Fig. 10.9(b)] > Default Model **Browse** > pick **clamp_arm.prt** [Fig. 10.9(c)] > **Preview** *on*

**Figure 10.9(a)** Type: Drawing

**Figure 10.9(b)** New Drawing Dialog

**Figure 10.9(c)** Preview clamp_arm.prt

Click: **Open** > Template **c_drawing** [Fig. 10.9(d)] > **OK** > CLAMP_ARM_DESIGN from the Select Instance dialog box [Fig. 10.9(e)] > **Open** [Fig. 10.9(f)] > 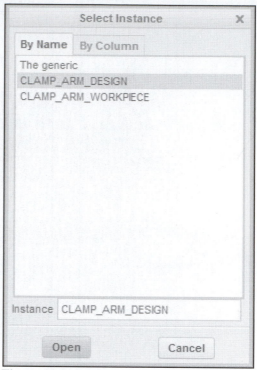 > ☐ (Select All) > **LMB** in the Graphics Window

**Figure 10.9(d)** Default Model clamp_arm.prt

**Figure 10.9(e)** Select Instance Dialog Box

**Figure 10.9(f)** Three Standard Views Displayed

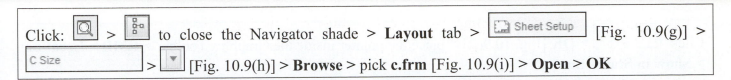

Click: [icon] > [icon] to close the Navigator shade > **Layout** tab > [Sheet Setup icon] Sheet Setup [Fig. 10.9(g)] >

[C Size] > [▼] [Fig. 10.9(h)] > **Browse** > pick **c.frm** [Fig. 10.9(i)] > **Open** > **OK**

**Figure 10.9(g)** Sheet Setup Dialog Box

**Figure 10.9(h)** Sheet Setup Dialog Box > Browse

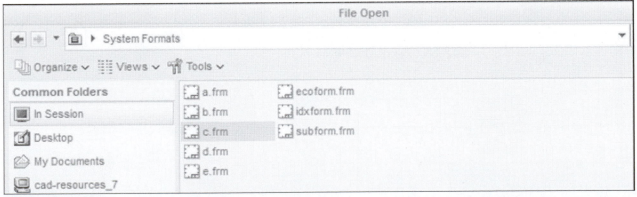

**Figure 10.9(i)** Open c.frm

451

Click:  > OK [Fig. 10.9(j)] > place the pointer inside the Graphics Toolbar > **RMB** > **Location** > **Show in Status Bar**

Since you created annotations of driving dimensions on the model, they will automatically display on the drawing along with the set datum tag annotations.

**Figure 10.9(j)** Drawing and Format

Press: **RMB > Insert General View** [Fig. 10.9(k)] > **OK** > pick a position for the new view [Fig. 10.9(l)]

**Figure 10.9(k)** Press RMB > Insert General View

**Figure 10.9(l)** New View Inserted

**Figure 10.9(m)** Default orientation: Isometric

**Figure 10.9(n)** New Isometric View

Click: **Annotate** tab > 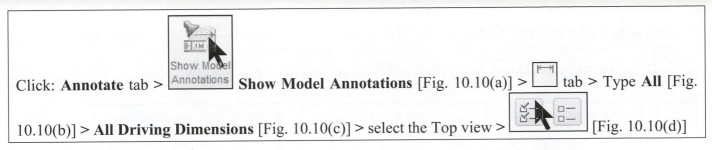 **Show Model Annotations** [Fig. 10.10(a)] > tab > Type **All** [Fig. 10.10(b)] > **All Driving Dimensions** [Fig. 10.10(c)] > select the Top view > [Fig. 10.10(d)]

**Figures 10.10(a-c)** Show Model Annotations Dialog Box (Dimensions tab) *(your list may appear differently)*

**Figure 10.10(d)** Select All Annotations *(your list may show more annotations)*

Click: **Apply** >  tab [Fig. 10.10(e)] > [Fig. 10.10(f)] *(your list may appear differently)* > tab
> [Fig. 10.10(g)] > **OK** > **LMB** to deselect [Fig. 10.10(h)] > **Ctrl+S** > **OK**

**Figure 10.10(e)** Model Datums Tab

**Figure 10.10(f)** Select All

**Figure 10.10(g)** Model Notes Tab > Select All

**Figure 10.10(h)** Dimensions, Axes, and Notes Displayed

Select the general view > press **RMB** > ☑ Lock View Movement (uncheck) > ⊕ hold down **LMB** and move the view [Fig. 10.11(a)] > release the **LMB** > **LMB** outside of the view to set the position [Fig. 10.11(b)]

**Figure 10.11(a)** Move View

**Figure 10.11(b)** New Position for General (Isometric) View

Press: **Ctrl+S** > **OK** > pick on the Right Side view > press **RMB** > **Delete** [Fig. 10.12(a)] > roll the (thumbwheel) **MMB** to zoom in on the Front view [Fig. 10.12(b)]

**Figure 10.12(a)** Press RMB > Delete the Right Side View

**Figure 10.12(b)** Front View

Pick on the hole dimension [Fig. 10.12(c)] > press **LMB** and move it to a new location > release the **LMB** > reposition dimensions as needed > reposition each datum tag by picking on the item and then moving it to a new location 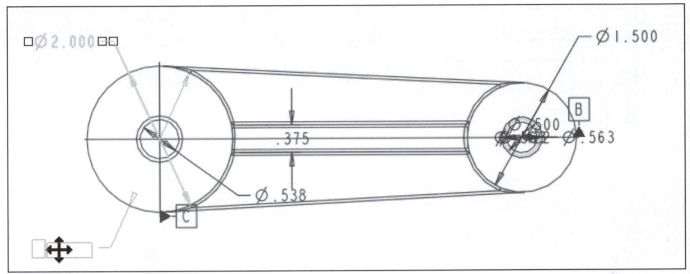 > to erase an item, pick on it > press **RMB** > **Erase** > **LMB** to accept > pick on the **2.000** diameter dimension > press **RMB** > **Flip Arrows** [Fig. 10.12(d)]

**Figure 10.12(c)** Move the Dimension

**Figure 10.12(d)** Flip Arrows

459

Pick on the note in the Top view > press **RMB** > [ Move Item to View ] [Fig. 10.13(a)] > select the Front view > press **LMB** on the note > move the pointer [Figs. 10.13(b-c)] > release the **LMB** > **LMB** to deselect

**Figure 10.13(a)** Press RMB > Move Item to View

**Figure 10.13(b)** Press LMB on the Note

**Figure**      **10.13(c)**      Move      the      Note

Click: [General ▼] (lower right-hand filter selection) > [Dimension ▼] > zoom in on the Front view > hold down the **Ctrl** key and select three dimensions of the hole [Fig. 10.13(d)] > release the **Ctrl** key > press **RMB** > **Erase** > move the pointer off of any entities > **LMB** to accept and deselect > [Dimension ▼] > [General ▼] > continue to move and erase dimensions to clean up the view > erased dimensions can be unerased using the Drawing Tree ( ⊞ *to open the Navigator shade* > *select the erased dimension* > *RMB* > *Unerase*) [Fig. 10.13(e)]

**Figure 10.13(d)** Erase Dimensions

**Figure 10.13(e)** Unerase Dimension

Move the **5.00** dimension from the Top view to the Front view > Delete dimensions that do not have a detailing purpose (dimension air- as with the 1.75 dimension). Deleted dimensions can be retrieved by using Show Model Annotations again. > Complete the drawing by erasing unneeded dimensions, flipping arrows, and repositioning dimensions, notes, and datum tags as needed [Fig. 10.13(f)]. > **LMB** to deselect > for the **.725** dimension, press **RMB** > **Dimension - New References** > select the small circular protrusion's vertical edge > move the pointer > **MMB** to place the dimension > **Return** from the Menu Manager > **LMB** in the Graphics Window > 🔍 > 💾 > **OK** > (**Ok** *if needed*) > **File** > **Manage File** > **Delete Old Versions** > **Enter** > **File** > **Save As** > Type ▾ > **Zip File (\*.zip)** > **OK** > (**Ok** *if needed*) > upload to your course interface or attach to an email and send to your instructor or yourself > **File** > **Close** *Though not necessarily complete or correct to ASME Y14.5M standards, we will leave this drawing as is. You may finalize this drawing after completing the next lesson.*

**Figure 10.13(f)** Completed Drawing

**Figure 10.14** Clamp Foot Drawing

## Creating a Drawing of the Clamp Foot

Before starting the drawing you will need to set your directory to the folder that contains the Swing Clamp Assembly (and therefore all the components of the assembly), and this time you will bring the Clamp_Foot (Fig. 10.14) in session by opening it. The standard three views of the model will display automatically on the drawing, since you will use the default drawing template for this project.

Click: **File > Manage Session > Select Working Directory** > select the directory where the **clamp_assembly.asm** was saved > **OK > File > Open > clamp_foot.prt > Open** (Fig. 10.15) > **File > Close** *(this keeps the part in session but not active or open)*

**Figure 10.15** Clamp_Foot

Press: **Ctrl+N** >  Drawing > Name **clamp_foot** *(keep the default settings)* > ☑ Use default template [Fig. 10.16(a)] > **OK** > Template **b_drawing** [Fig. 10.16(b)] > make sure that the Default Model is the **CLAMP_FOOT.PRT** *(if it is not, change it with **Browse**)* > **OK** [Fig. 10.16(c)] > **Ctrl+S** > **OK**

**Figure 10.16(a)** Use Default Template

**Figure 10.16(b)** B-Size. CLAMP_FOOT.PRT is Default Model

**Figure 10.16(c)** Clamp Foot Drawing

**Figure 10.16(d)** Clamp Foot Drawing Sheet Setup

**Figure 10.16(e)** Sheet Size

Pick on the Right Side view > press **RMB** > **Delete** >  > ☐ (Select All) (unchecked) > **LMB** > **Layout** tab > *off* > **Annotate** tab > > tab > Type **All** > with the **Ctrl** key pressed, select the Top and Front views > release the **Ctrl** key > > uncheck the two **360** dimensions *(as you move the mouse pointer over the entries in the Show or Type column, the corresponding dimension highlights in the Graphics Window)* [Fig. 10.16(f)] > **Apply** > tab > > **OK** > **LMB** to deselect

**Figure 10.16(f)** Show Model Annotations

466

Change the datums display to ASME, click: **File** > **Prepare** > **Drawing Properties** > Detail Options **change** > Option: **gtol_datums** > Value: [ ▼ ] > **std_asme** > **Add/Change** > **OK** > **Close** > Filter: [General ▼] > [Dimension ▼] > window-in to capture all the dimensions > press **RMB** > **Cleanup Dimensions** [Fig. 10.16(g)] > [☐ Create Snap Lines] [Fig. 10.16(h)] > Increment **.5** > **Apply** > [ ⊡ ] > **Close** > **LMB** > **Ctrl+S** > **OK** > Filter: [Dimension ▼] > [General ▼]

**Figure 10.16(g)** Press RMB > Cleanup Dimensions

**Figure 10.16(h)** Clean Dimensions Dialog Box

467

Click:  [Fig. 10.16(i)] > **In Session** > **clamp_foot.prt** > **Open** [Fig. 10.16(j)] > **View Manager** > **Sections** tab > **New** > **Planar** > *type* **A** [Fig. 10.16(k)] > **Enter** > pick on datum **C** [Fig. 10.16(l)] >  >  > 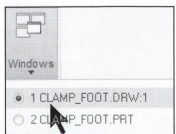 > **RMB** on **A** > uncheck **Show Section** > **Close** > **View** tab > **Windows** > **CLAMP_FOOT.DRW:1** [Fig. 10.16(m)] > **Ctrl+S** > **Enter**

**Figure 10.16(i)** Open the Clamp_Foot from In Session

**Figure 10.16(j)** Part Window Activated

**Figure 10.16(k)** View Manager Dialog Box

**Figure 10.16(l)** Part Xsection A

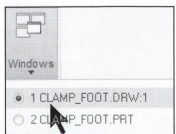

**Figure 10.16(m)** Drawing Active

468

**Figure 10.16(n)** Press RMB > Properties

**Figure 10.16(o)** Section View

469

Click: **Annotate** tab > Cleanup the drawing views by erasing unneeded dimensions, moving datums, and moving dimensions to the appropriate view [Fig. 10.16(p)]. > in the Graphics Window, **LMB** to deselect >  **Refit** > **Ctrl+S** > **Enter** > **File** > **Manage File** > **Delete Old Versions** > **Enter** > **File** > **Save As** > Type > **Zip File (*.zip)** > **OK** > upload to your course interface or attach to an email and send to your instructor or yourself > **File** > **Close** > **File** > **Close**

**Figure 10.16(p)** Section View and Front View Cleaned Up

Though not complete, or completely correct to ASME Y14.5M standards, we will leave this drawing as is. You may finalize the drawing after completing the detail drawing in the next lesson.

**Figure 10.17** Clamp Ball Drawing

## Creating a Drawing of the Clamp Ball

Before starting the Clamp Ball drawing (Fig. 10.17) for the Clamp Ball part (Fig. 10.18), you will build a template that can be used to create any drawing.

Click: **File > Manage Session > Select Working Directory** > select the directory where the **clamp_assembly.asm** was saved > **OK** > 📄 **Create a new model** > ⊙ 🖳 Drawing > Name: **detail_template** > ☐ Use default template [Fig. 10.19(a)] > **OK** > Default Model **Browse** > pick **clamp_ball.prt > Open** > ⊙ Empty with format > Format **Browse** > **a.frm > Open** [Fig. 10.19(b)] > **OK**

**Figure 10.18** Clamp_Ball

**Figure 10.19(a)** New Dialog Box

**Figure 10.19(b)** New Drawing Dialog Box

Click: **Tools** tab > Template **Drawing template mode** > 🔍 > **Layout** tab > Template View [Fig. 10.19(c)]

**Figure 10.19(c)** Tools Tab > Template > Layout Tab > Template View

The Template View Instructions dialog box opens, press **LMB** on the Title bar of the dialog box > move the pointer (make sure the box hangs over the Graphics Window edge) [Fig. 10.19(d)] > release the **LMB**

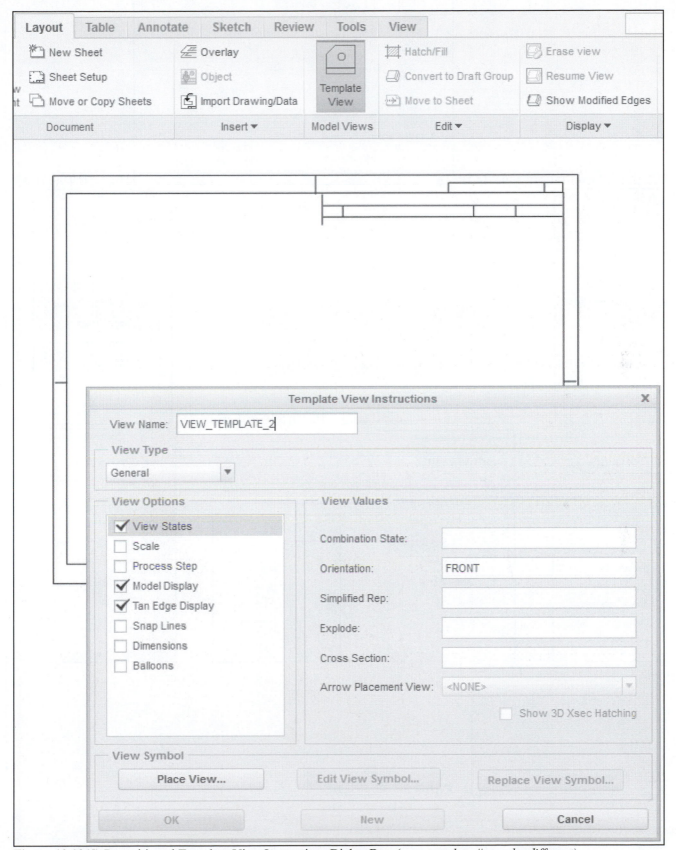

**Figure 10.19(d)** Repositioned Template View Instructions Dialog Box *(your template # may be different)*

**View Values**

Combination State: [ ]

Orientation: TOP

Simplified Rep: [ ]

Explode: [ ]

Cross Section: [ ]

Arrow Placement View: <NONE> ▾

☐ Show 3D Xsec Hatching

View Values, Orientation: *type* **TOP** [Fig. 10.19(e)] >

**View Options**
- ☑ View States
- ☑ Scale
- ☐ Process Step
- ☑ Model Display
- ☑ Tan Edge Display
- ☐ Snap Lines
- ☐ Dimensions
- ☐ Balloons

**View Options**
- ☑ View States
- ☑ Scale
- ☐ Process Step
- ☑ Model Display
- ☑ Tan Edge Display
- ☐ Snap Lines
- ☐ Dimensions
- ☐ Balloons

>

View Scale: 2.000000

[Fig. 10.19(f)] > **Enter** >

**View Options**
- ☑ View States
- ☑ Scale
- ☐ Process Step
- ☑ Model Display
- ☑ Tan Edge Display
- ☐ Snap Lines
- ☐ Dimensions
- ☐ Balloons

>

**View Values**
- ○ Follow Environment
- ○ Wireframe
- ● Hidden Line
- ○ No Hidden
- ○ Shading

**View Values**
- ○ Tan Solid
- ● No Disp Tan
- ○ Tan Ctrln
- ○ Tan Phantom
- ○ Tan Dimmed
- ○ Tan Default

**View Options**
- ☑ View States
- ☑ Scale
- ☐ Process Step
- ☑ Model Display
- ☑ Tan Edge Display
- ☐ Snap Lines
- ☑ Dimensions
- ☐ Balloons

>

> > >

**View Values**
- ☐ Create Snap Lines
- Incremental Spacing: 0.500000
- Initial Offset: 0.500000
- Set Display Priorities...

> **Enter** > ☑ View States

**Figure 10.19(e)** View States

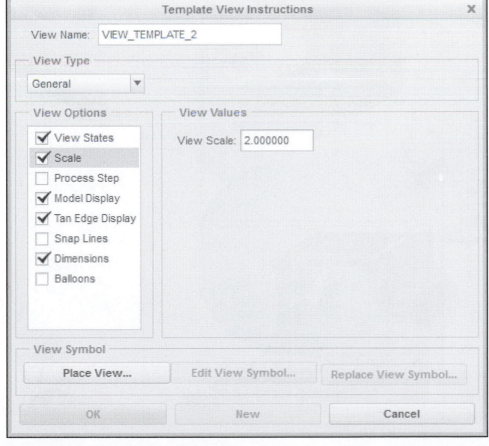

**Figure 10.19(f)** View Scale

Click: [Place View...] > pick a position for the Top view [Fig. 10.19(g)] > **OK** to close the Template View Instructions dialog box > **Lock View Movement** *on* > **Ctrl+S** > **OK** > **File** > **Close** > **File** > **Manage Session** > **Erase Not Displayed** [Fig. 10.19(h)] > **OK** > **File** > **Open** > **clamp_ball.prt** > **Open** [Fig. 10.19(i)] > **File** > **Close** (clamp_ball is active and in session but not displayed in a window)

**Figure 10.19(g)** Place the View *(the Template View Instructions Dialog Box may be behind the Creo 2.0 Graphics Window)*

**Figure 10.19(h)** Erase Not Displayed

**Figure 10.19(i)** clamp_ball.prt

Click: 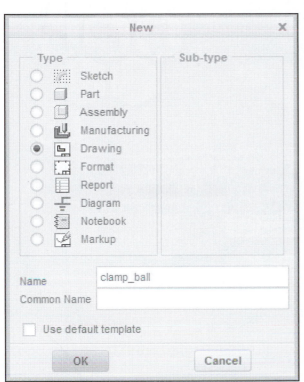 **Create a new model** > ⦿ 🖳 Drawing > Name **clamp_ball** > ☐ Use default template [Fig. 10.20(a)] > **OK** > ⦿ Use template > Template **Browse** > File name: detail_template.drw > **Open** [Fig. 10.20(b)] > **OK** > pick on the view > press **RMB** > ☐ Lock View Movement (uncheck) > **LMB** on the view and move it to a better position > release the **LMB** [Fig. 10.20(c)] > 🔍 > **Ctrl+S** > **MMB** > **LMB** to deselect

**Figure 10.20(a)** New Drawing

**Figure 10.20(b)** Template: detail_template

**Figure 10.20(c)** Repositioned View

477

Click:  (uncheck) > **LMB** > **Annotate** tab > **Show Model Annotations** > select the
view > tab > > **OK** > cleanup the drawing view by erasing
unneeded dimensions and the SCALE note, moving datums, etc. [Fig. 10.20(d)] > **LMB** to deselect

**Figure 10.20(d)** One-view Drawing

478

Click: 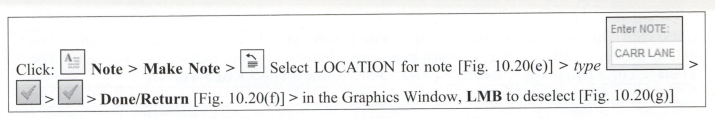 **Note > Make Note >** Select LOCATION for note [Fig. 10.20(e)] > *type* > > **Done/Return** [Fig. 10.20(f)] > in the Graphics Window, **LMB** to deselect [Fig. 10.20(g)]

**Figure 10.20(e)** Pick a Position for the Note

**Figure 10.20(f)** CARR LANE Note

**Figure 10.20(g)** Completed Note

**Figure 10.20(h)** Completed Drawing

You can create drawings of the remaining clamp assembly parts. And, a variety of part, assembly, and drawing projects can be downloaded from the website ***www.cad-resources.com***.

**Figure 11.1(a)** Anchor Drawing

## OBJECTIVES

- Establish a **Drawing Options** file to use when detailing
- Identify the need for **views** to clarify interior features of a part
- Create **Cross Sections** using datum planes
- Produce **Auxiliary Views**
- Create **Detail Views**
- Use **multiple drawing sheets**
- Apply **standard drafting conventions** and line types to illustrate interior features

## REFERENCES AND RESOURCES

For **Resources** go to www.cad-resources.com > click on the Creo Parametric 2.0 Book cover

- Lesson 11 Lecture
- Lesson 11 3D models embedded in a PDF
- Book Projects PDF
- Project Lectures
- Creo Parametric 2.0 Quick Reference Card
- Creo Parametric 2.0 Configuration Options

## PART DRAWINGS

Designers and drafters use drawings to convey design and manufacturing information. Drawings consist of a **Format** and views of a part (or assembly). Standard views, sectional views, detail views, and auxiliary views are utilized to describe the objects' features and sizes [Fig. 11.1(a)]. **Sectional views**, also called **sections**, are employed to clarify and dimension the internal construction of an object. Sections are needed for interior features that cannot be clearly described by hidden lines in conventional views. **Auxiliary views** are used to show the *true shape/size* of a feature or the relationship of features that are not parallel to any of the principal planes of projection. Many objects have inclined surfaces and features that cannot be adequately displayed and described by using principal views alone. To provide a clearer description of these features, it is necessary to draw a view that will show the *true shape/size*.

**Figure 11.1(b)** Anchor Drawing, Sheet 2

## Anchor Drawing

You will be creating a multiple sheet detail drawing of the Anchor [Fig. 11.1(b)]. The front view will be a full section. A right side view and an auxiliary view are required to detail the part. Views will be displayed according to visibility requirements per ASME standards, such as no hidden lines in sections. The part is to be dimensioned according to ASME Y14.5M. You will add a standard format. Detailed views of other parts will be introduced to show the wide variety of view capabilities.

Launch **Creo Parametric 2.0 > Select Working Directory** > select the directory where the **anchor.prt** was saved > **OK** > [icon] > [◉ Drawing] > Name **anchor** > [☐ Use default template] *(if "Use default template" is checked; the Front, Top, and Right views will be automatically created)* [Fig. 11.2(a)] > **OK** > Default Model **Browse** [Fig. 11.2(b)] > pick **anchor.prt** [Fig. 11.2(c)] > **Preview** *on* > **Open**

**Figure 11.2(a)** New Drawing

**Figure 11.2(b)** New Drawing Dialog Box

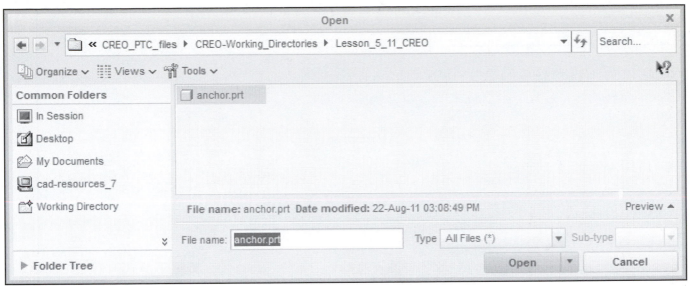

**Figure 11.2(c)** Anchor Part

Next to Standard Size, click: [☑️] > pick **D** [Fig. 11.2(d)] > **OK** > press **RMB** > **Sheet Setup** > [D Size ▼] [Fig. 11.2(e)] > [☑️] > **Browse** (System Formats Open dialog box opens) > pick **d.frm** [Fig. 11.2(f)] > **Open** > **OK** [Fig. 11.2(g)]

**Figure 11.2(d)** Standard Size D

**Figure 11.2(e)** Sheet Setup

**Figure 11.2(f)** System Formats

**Figure 11.2(g)** "D" Size Format

Click: **File > Prepare > Drawing Properties >** Detail Options **change** (the Options dialog box opens) [Fig. 11.2(h)] > make changes to create a new **.dtl** file *(or* 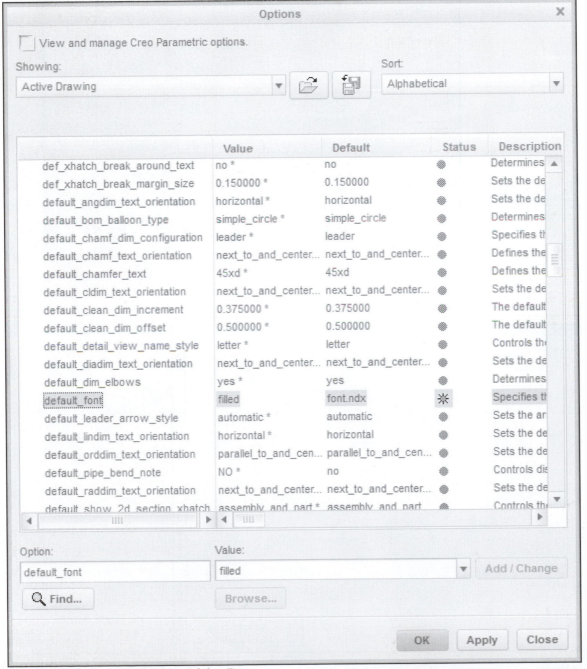 *> pick a previously saved **.dtl** from your directory list > Open) >* Change the following options to the values listed below:

Option:   *text_height*           Value:   *.25*      >      **Add/Change**
Option:   *default_font*          Value:   *filled*   >      **Add/Change**
Option:   *arrow_style*           Value:   *filled*   >      **Add/Change**
Option:   *allow_3d_dimensions*   Value:   *yes*      >      **Add/Change**

Click: **Apply** > Sort: [▼] > **Alphabetical** [Fig. 11.2(h)] > [icon] **Save a copy of the currently displayed configuration file** > *type (a unique name for your file)* **draw_options** > **OK** > **Close** > **Close**

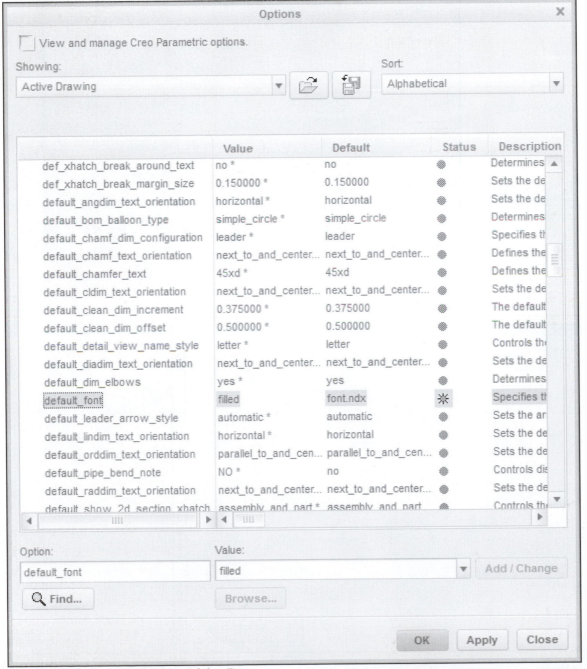

**Figure 11.2(h)** Drawing Options Dialog Box

485

Click: [icon] (to close the Navigator so that you can see a larger drawing) > [icon] > [icon] **Hidden Line** >

**Layout** tab > [General] **Create a general view** > **No Combined State** (highlighted) > **OK** > pick a position for the view [Fig. 11.3(a)] > Model view names **FRONT** > **OK** > **LMB** to deselect [Fig. 11.3(b)]

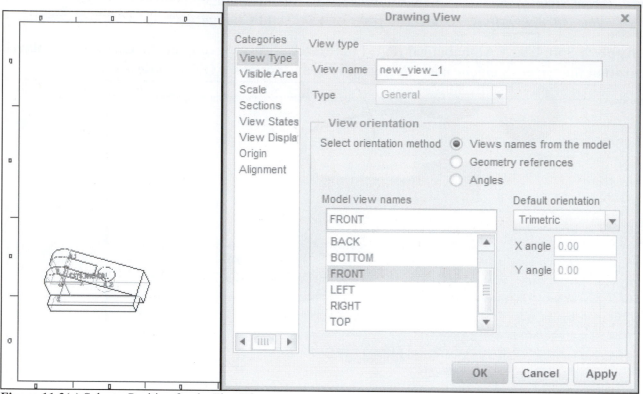

**Figure 11.3(a)** Select a Position for the First View

**Figure 11.3(b)** Front View

Click: [icon] **Datum Display Filters** > (**Select All**) *off* > **LMB** in the Graphics Window > **Ctrl+R** > add two more views; pick on the Front view (highlights) > [Projection] > [Select CENTER POINT for drawing view.] pick a position for the Right Side view [Fig. 11.3(c)] > move the mouse pointer off of the Right Side view > **LMB** to deselect > pick on the Front view > press **RMB** > **Insert Projection View** > [Select CENTER POINT for drawing view.] pick a position for the Top view > move the mouse pointer off of the view > **LMB** to deselect [Fig. 11.3(d)] > from the Ribbon, **Lock View Movement** *off* > pick on a view, hold down the **LMB**, and reposition as needed > release the **LMB** > move the mouse pointer off of the view > **LMB** to deselect > [icon] > [icon] > **OK** [Fig. 11.3(e)]

**Figure 11.3(c)** Add Right Side Projected View

**Figure 11.3(d)** Add Top Projected View

**Figure 11.3(e)** Repositioned Views

The top view does not help in the description of the part's geometry. Delete the top view before adding an auxiliary view that will display the true shape of the angled surface.

Pick on the Top view > press **RMB** [Fig. 11.3(f)] > **Delete** > [🔍] > **Ctrl+S** > **OK**

**Figure 11.3(f)** Select the Top View > Press RMB > Delete

Click: **Auxiliary** [Fig. 11.3(g)] > 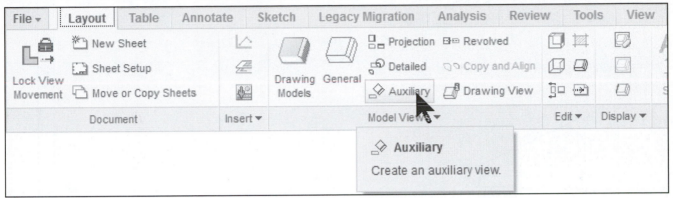 ⇨ Select edge of or axis through, or datum plane as, front surface on main view. pick on datum **E** [Fig. 11.3(h)] > move the mouse pointer towards the upper right [Fig. 11.3(i)] > ⇨ Select CENTER POINT for drawing view. **LMB** to place the view [Fig. 11.3(j)] > move the mouse pointer off of the view > **LMB** to deselect > 🔍 > 💾 > **MMB**

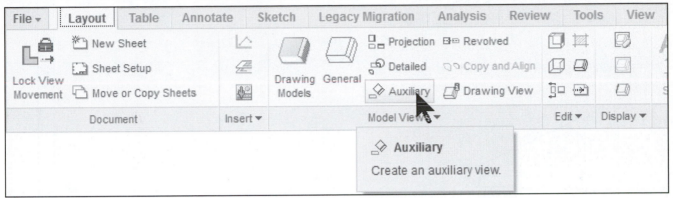

**Figure 11.3(g)** Create an Auxiliary View

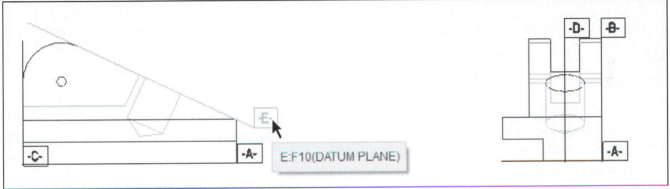

**Figure 11.3(h)** Select Datum Plane E or the Angled Edge

**Figure 11.3(i)** Move the Mouse Pointer to a new Location to Place the View

**Figure 11.3(j)** Auxiliary View

The correct standard was not selected when you made changes to the Drawing Options. ASME symbols for datum planes (gtol_datums) are the correct style standard used on drawings. The ANSI style [Fig. 11.3(k)] was discontinued in 1994, though retained by some companies as an "in house" standard. For outside vendors and for manufacturing internationally, the ISO-ASME standards (ASME Y14.5) should be applied to all manufacturing drawings.

**Figure 11.3(k)** gtol_ansi

**Figure 11.3(l)** gtol_asme

Next, change the Front view into a sectional view. Section **A** was created in the Part mode. Pick on the Front view as the view to be modified [Fig. 11.4(a)] > press **RMB** > **Properties** [Fig. 11.4(b)] > Categories **Sections** > ⦿ 2D cross-section > ➕ **Add cross-section to view** > ✓ A from the Name list [Fig. 11.4(c)] > **Apply** > **Close** [Fig. 11.4(d)] > in the Graphics Window, **LMB** to deselect

**Figure 11.4(a)** Select the Front View

**Figure 11.4(b)** RMB > Properties

**Figure 11.4(c)** Drawing View Dialog Box

**Figure 11.4(d)** Section A-A

491

Show all dimensions and axes (centerlines), click: **Annotate** tab > [icon] > [icon] Dimensions tab > with the **Ctrl** key pressed, pick on the Front and the Auxiliary views [Fig. 11.5(a)] > release the **Ctrl** key > [icon] > **Apply** > [icon] Datums tab > with the **Ctrl** key pressed, select the Right view > release the **Ctrl** key > [icon] [Fig. 11.5(b)] > **Apply** > **Cancel**

**Figure 11.5(a)** Show Model Annotations Dialog Box

**Figure 11.5(b)** Dimensions and Axes displayed in the three views

Select all dimensions by enclosing them in a selection box > press **RMB** > **Cleanup Dimensions** [Fig. 11.5(c)] > ☐ Create Snap Lines > Increment .5 > **Enter** > **Apply** [Fig. 11.5(d)] > **Close** > in the Graphics Window, **LMB** to deselect > 🔍

**Figure 11.5(c)** Press RMB > Cleanup Dimensions

**Figure 11.5(d)** Cleaned Dimensions and Clean Dimensions Dialog Box *(your drawing may look different)*

Pick on the **1.00** diameter dimension > press **RMB** > **Move Item to View** [Fig. 11.6(a)] > pick on the Auxiliary view [Fig. 11.6(b)] > pick on and reposition the **1.00** dimension [Fig. 11.6(c)] > in each view, reposition dimensions, move dimension text, clip extension lines, or flip arrows where appropriate > reposition, move, clip, or erase axes and position datums > pick the Front view > press **RMB** > **Add Arrows** > pick on the Auxiliary view > move the pointer off of the view > **LMB** to deselect [Fig. 11.6(d)]

**Figure 11.6(a)** Select the **1.00** Diameter Dimension

**Figure 11.6(b)** Diameter Dimension **1.00** Moved to Auxiliary View

**Figure 11.6(c)** Repositioned **1.00** Diameter Dimension

**Figure 11.6(d)** Edited Auxiliary View

Add a reference dimension to the small hole, click: [icon] **Reference Dimension -New References** > pick the horizontal axis of the small hole > pick datum **A** [Fig. 11.7(a)] > **MMB** to place the reference dimension [Fig. 11.7(b)] > **MMB** (or **Return** from the Menu Manager) > move the pointer off of the dimension > **LMB** to deselect

**Figure 11.7(a)** Select Datum A

**Figure 11.7(b)** Placed Reference Dimension

To change the number of decimal places for a dimension > pick the dimension(s) > press **RMB** > **Properties** > Decimal Places **3** > **OK** > **LMB** to deselect > press the **Ctrl** key, select the Auxiliary and Right views > release the **Ctrl** key > press **RMB** > **Properties** > Display style ⬛ > **No Hidden** > **OK** > **LMB** to deselect (Fig. 11.8)

**Figure 11.8** Edited Right Side View

Add the edges of the small through hole back into the auxiliary view. The edges will display in the graphics window in a lighter color, but print as dashed on the drawing plot.

Pick the Auxiliary view > **Layout** tab > 🔲 Edge Display > **Hidden Line** > press and hold the **Ctrl** key and pick near where the small through hole would show as hidden in the Auxiliary view [Fig. 11.9(a)] > pick the opposite edge [Fig. 11.9(b)] > release the **Ctrl** key > **OK** > **Done** > ⬛ > 🖫 > **OK** [Fig. 11.9(c)]

**Figure 11.9(a)** Select near the Hole Edge

**Figure 11.9(b)** Select near the Opposite Edge of the Hole

**Figure 11.9(c)** Small Through Hole's Edge Lines Displayed

To increase the functionality of the drawing, you will need to master a number of capabilities. Partial views, detail views, using multiple sheets, and modifying section lining (crosshatch lines) are just a few of the many options available in Drawing Mode. Create a Detail View.

Click: [⊕ Detailed] > [➡ Select center point for detail on an existing view.] pick the top edge of the hole in the Front section view [Fig. 11.10(a)] > [➡ Create points for the spline to pass through.] *each **LMB** pick adds a point to the spline* > **MMB** to end (and close) the spline [Fig. 11.10(b)] *(you do not have to "close" the spline sketch – just get near the start of the spline and clicking the MMB will close the spline for you)*

**Figure 11.10(a)** Select Center Point for Detail View

**Figure 11.10(b)** Sketch a Spline (MMB to close the spline)

497

pick in the upper right corner of the drawing sheet to place the view [Fig. 11.10(c)] > with the Detailed view selected, press: **RMB > Properties** [Fig. 11.10(d)] > Categories **Scale >** ⦿ Custom scale    2.000    **> 1.500** [Fig. 11.10(e)] **> Enter > Apply > Close > LMB** to deselect **> Ctrl+S > OK**

**Figure 11.10(c)** Detail View A

**Figure 11.10(d)** Drawing View Dialog Box

**Figure 11.10(e)** Custom Scale

498

Add an axis to the hole in view DETAIL A; click **Annotate** tab >  [Fig. 11.10(f)] > pick on the hole (feature) in view DETAIL A [Fig. 11.10(g)] > **Apply** [Fig. 11.10(h)] > **Cancel** > > > **Ctrl+S** > **Enter**

**Figure 11.10(f)** Datums

**Figure 11.10(g)** Select the Hole

**Figure 11.10(h)** Show Axis

In DETAIL A, erase datum **A,** clip datum **E** and clip the axis. > with the **Ctrl** key pressed, pick on the text items: **SECTION A-A, SEE DETAIL A** and **DETAIL A SCALE 1.500** [Fig. 11.11(a)] > release the **Ctrl** key > press **RMB** > **Text Style** > next to Height ☐ Default (uncheck) > Height **.375** > **Enter** [Fig. 11.11(b)] > **Apply** > **OK** [Fig. 11.11(c)] > **LMB** to deselect

**Figure 11.11(a)** Select the Text Items and Change Their Height

**Figure 11.11(b)** Text Style Dialog Box

500

**Figure 11.11(c)** Changed Text Style (Height) *(your drawing may appear differently)*

Change the height of the section identification lettering to **.375**, pick on "A" > press **RMB** > **Text Style** [Fig. 11.11(d)] > next to Height ☐ Default (uncheck) > Height **.375** > **Enter** > **Apply** [Fig. 11.11(e)] > **OK** > **LMB** > window-in all drawing text and dimensions > press **RMB** > **Text Style** > next to Font ☐ Default (uncheck) > ▼ > Font ☐font > **Apply** > **Close** > **LMB** to deselect

**Figure 11.11(d)** Text Style of the Section Identification Lettering

**Figure 11.11(e)** New Text Height

501

The shape and size of the Arrows for the cutting plane and dimensions are too small. These are controlled by the Drawing Options, click: **File** > **Prepare** > **Drawing Properties** > Detail Options **change** > *draw_arrow_length .25* > **Enter** > *draw_arrow_width .10* > **Enter** > *crossec_arrow_length .375* > **Enter** > *crossec_arrow_width .125* > **Enter** > *default_font* > **default** [Fig. 11.12(a)] > **Add / Change** > *arrow_style* > *filled* > **Enter** > **Apply** > **Close** > **Close** > scroll the **MMB** to zoom in [Fig. 11.12(b)] > 🔍 > **Ctrl+S** > **OK**

| | | | | |
|---|---|---|---|---|
| create_area_unfold_segmented | yes * | yes | ⬤ | Makes the display of dimensions in are |
| crossec_arrow_length | 0.375000 | 0.187500 | ⬤ | Sets the length of the arrow head on th |
| crossec_arrow_style | tail_online * | tail_online | ⬤ | Determines which end of cross-sectio |
| crossec_arrow_width | 0.125000 | 0.062500 | ⬤ | Sets the width of the arrow head on th |
| crossec_text_place | after_head * | after_head | ⬤ | Sets the location of cross-section text |

**Figure 11.12(a)** Cross Section Drawing Options

**Figure 11.12(b)** Section Arrows Length and Width Changed

Erase the **118.0** angle dimension from the Front view. > Since the model and the drawing are associative, make a modification to the slot dimension. Pick on the dimension shown in Figure 11.12(c). > press **RMB** > **Modify Nominal Value** > *type* **3.125** [Fig. 11.12(d)] > **Enter** > **LMB** to deselect (dimension changes color) > **Review** tab > [🔄 Regenerate Active Model] > reposition dimensions, datums and notes [Fig. 11.12(e)]

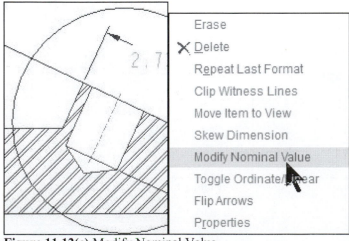

**Figure 11.12(c)** Modify Nominal Value

**Figure 11.12(d)** Modify to **3.125**

**Figure 11.12(e)** Regenerated Part and Drawing

Pick on the **1.125** dimension in the Front section view [Fig. 11.13(a)] > press **RMB** > **Move Item to View** > pick on view DETAIL A [Fig. 11.13(b)]

Erase
✕ Delete
Repeat Last Format
Clip Witness Lines
Move Item to View
Skew Dimension
Modify Nominal Value
Toggle Ordinate/Linear
Flip Arrows
Properties

**Figure 11.13(a)** Select the **1.125** Hole Depth Dimension

Zoom in on the Front view >  (to open the Navigator) > **Annotate** tab > expand the Drawing Tree ▼ 📖 new_view_1 > ▼ Annotations > Model: d24 [Fig. 11.13(c)] > **RMB** > **Unerase** [Fig. 11.13(d)] > **RMB** > **Move Item to View** > pick on view DETAIL A > reposition and clip the moved dimensions as needed and change the decimal places of the **118** degree dimension to **0** > 🔍 > 🖼

**Figure 11.13(b)** Move Dimension to DETAIL A

**Figure 11.13(c)** Drawing Tree

**Figure 11.13(d)** Unerase the Angle Dimension [your view number and dimension (d) number may be different]

**Figure 11.14(a)** Detail A

**Figure 11.14(b)** Right Side View

**Figure 11.14(c)** Section A-A

505

Next, you will change the boundary of view DETAIL A. Click: **Layout** tab > pick on view DETAIL A > slightly move the mouse pointer > press **RMB** (if you press RMB outside the view outline, you get a pop-up list of options [Fig. 11.15(a)], whereas if you press RMB inside the view outline, there are more options) [Fig. 11.15(b)] > **Properties** > click the **Spline defined** button [Fig. 11.15(c)] > sketch the spline again in the Front (SECTION A-A) view [Fig. 11.15(d)] > **MMB** to end spline > **Apply**

**Figure 11.15(a)** RMB Outside

**Figure 11.15(b)** RMB Inside

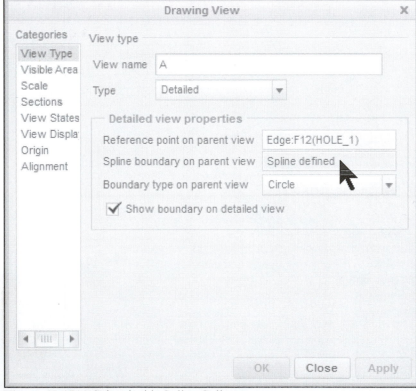

**Figure 11.15(c)** Select inside Spline Collector

**Figure 11.15(d)** Sketch the Spline Again

**Figure 11.15(e)** New Circle Position

**Figure 11.15(f)** Updated DETAIL A

**Figure 11.15(g)** Anchor Drawing

Click: **Annotate** tab > pick on the **2.120 REF** dimension > press **RMB** > **Properties** > Decimal Places **3** > **Display** tab > remove the **REF** Suffix Suffix [        ] > add parentheses to the default text (@D) [Fig. 11.16(a)] > **OK** > **LMB** to deselect [Fig. 11.16(b)] > 🔍 > 🖉 > **Ctrl+S** > **Enter**

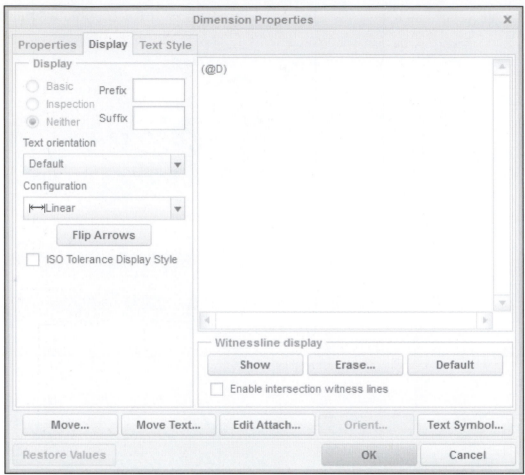

**Figure 11.16(a)** Dimension Properties Dialog Box

**Figure 11.16(b)** Correct ASME Standard Reference Dimension Style

Change the spacing and the angle of the section lining. Click: **Layout** tab > double-click on the Xhatching in SECTION A-A [Fig. 11.17(a)] > **Delete line** > **Angle** > **30** > **Spacing** > **Double** > **Half** > **Done** [Fig. 11.17(b)] > **LMB** to deselect >  [Fig. 11.17(c)] >  > **File** > **Save** > **OK**

**Figure 11.17(a)** Xhatching Properties

**Figure 11.17(b)** Xhatching at **30** Degree Angle

**Figure 11.17(c)** Xhatching

Change the spacing and the angle of the section lining in DETAIL A. Double-click on the Xhatching in view DETAIL A [Fig. 11.18(a)] > **Det Indep** > **Hatch** > **Spacing** > **Double** > **Angle** > **45** > **Done** [Fig. 11.18(b)] > **LMB** to deselect > **Review** tab > **Update Sheet** from the Ribbon >  > > > **OK**

**Figure 11.18(a)** Detail Xhatching Properties

**Figure 11.18(b)** Xhatching at 45 Degree Angle

Change the text style used on the drawing. Click: **Annotate** tab > enclose all of the drawing text and dimensions with a selection box [Fig. 11.19(a)] > press **RMB** > **Text Style**

**Figure 11.19(a)** Select all Text and Dimensions

Click: Font **Blueprint MT** [Fig. 11.19(b)] > **Apply** > **OK** > **LMB** to deselect > change the number of decimal places for dimensions as required [Fig. 11.19(c)] >  >  >  > **MMB**

**Figure 11.19(b)** Character- Font- Blueprint MT

**Figure 11.19(c)** New Text Style

Add a geometric tolerance to the angled surface. Click: [🔲 1M] **Create geometric tolerances** > [◺]
**Angularity** > Type **Datum** [Fig. 11.20(a)] > pick on datum **E** [Fig. 11.20(b)] > **Datum Refs** tab >
**Primary** tab > Basic [▶] > pick on datum **A** in the Front view as the Primary Reference Datum [Fig.
11.20(c)] > **Tol Value** tab > [✓ Overall Tolerance | 0.005] [Fig. 11.20(d)] > **Enter** > **OK** > in the Graphics
Window, **LMB** to deselect [Fig. 11.20(e)] > **Ctrl+S** > **OK**

**Figure 11.20(a)** Angularity, Type Datum

**Figure 11.20(b)** Pick Datum E

**Figure 11.20(c)** Primary Reference Datum A

**Figure 11.20(d)** Overall Tolerance Value **.005**

**Figure 11.20(e)** Geometric Tolerance

Create a *second sheet* with an isometric view of the Anchor. Click: **File > Options > Model Display >** Default model orientation **Isometric > OK > No > Layout** tab > [ New Sheet ] > [ Sheet Setup ] > [ D ] > [ D ▼ ] > **Browse** [Fig. 11.21(a)] > System Formats **c.frm** [Fig. 11.21(b)] > **Open > OK >** [🔍] > **Ctrl+S > OK**

**Figure 11.21(a)** Sheet Setup Dialog Box

**Figure 11.21(b)** c.frm

513

Click: General > **OK** > ⇨ Select CENTER POINT for drawing view. > **View Display** > Display style **No Hidden** > **Scale** > ⦿ Custom scale 1.500 [Fig. 11.21(c)] > **Enter** > **Apply** > **Close** > **LMB** [Fig. 11.21(d)]

**Figure 11.21(c)** Drawing View Custom Scale

**Figure 11.21(d)** C Format and Isometric View

Double-click on the pictorial view > **View Display** > Tangent edges display style **None** [Fig. 11.21(e)] > **OK** > **LMB** to deselect > Sheet 1 [Fig. 11.21(f)] > 🔍 > 💾 > **MMB**

**Figure 11.21(e)** Drawing View Dialog Box

**Figure 11.21(f)** Anchor Drawing Sheet 1

515

Click: [icon] **Open an existing model** > [icon] System Formats > [icon] d.frm > **Open** > **File** > **Save As** > **Save a Copy** > *type a unique name for your format:* New Name **DETAIL_FORMAT_D** > **OK** > **File** > **Close** > [icon] > pick **detail_format_d.frm** > **Open** > **File** > **Prepare** > **Drawing Properties** > **change** > Sort: **Alphabetical** > *text_height  .25* > **Add/Change** > *default_font  filled* > **Add/Change** > *arrow_style filled* > **Add/Change** > *draw_arrow_length  .25* > **Add/Change** > *draw_arrow_width  .08* > **Add/Change** [Fig. 11.22(a)] > **Apply** > **Close** > **Close** > **Ctrl+S** > **OK**

The **format** will have a **.dtl** associated with it, and the **drawing** will have a different **.dtl** file associated with it. *They are separate .dtl files.* When you activate a drawing and then add a format, the **.dtl** for the format controls the font, etc. for the format only.

| draw_arrow_length | .25 | 0.187500 | ● | Sets length of leader line arrows. |
|---|---|---|---|---|
| draw_arrow_width | .08 | 0.062500 | ● | Sets width of leader line arrows. Drives these: "draw_attac |
| draw_attach_sym_height | DEFAULT * | default | ● | Sets height of leader line slashes, integral signs, and boxes |
| draw_attach_sym_width | DEFAULT * | default | ● | Sets width of leader line slashes, integral signs, and boxes. |
| draw_dot_diameter | DEFAULT * | default | ● | Sets diameter of leader line dots. If set to "default," uses val |
| drawing_units | inch | | ☒ | Sets units for all drawing parameters. |
| leader_elbow_length | 0.250000 * | 0.250000 | ● | Determines length of leader elbow (the horizontal leg attach |
| line_style_standard | std_ansi * | std_ansi | ● | Controls text color in drawings. Unless set to "std_ansi," all |
| node_radius | DEFAULT * | default | ● | Sets the size of the nodes displayed in symbols. |
| sort_method_in_region | delimited * | delimited | ● | Determines repeat regions sort mechanism. String_only - so |
| sym_flip_rotated_text | no * | no | ● | If set to "yes," then for new symbol definitions which allow |
| text_height | .25 | 0.156250 | ● | Sets the default text height for newly created annotations. |

**Figure 11.22(a)** Drawing Option **.dtl** File for Format

*If you miss any item in this sequence it will not work. The text you will type is case sensitive. Press every "Enter". It works perfectly; but if you do not input it as shown, it will not display.*

Zoom into the title block region. Create notes for the title text and parameter text. Click: **File** > **Options** > **Sketcher** > [☑ Snap to grid] > **OK** > **No** > **Sketch** tab > **Draft Grid** > **Show Grid** > **Grid Params** > **X&Y Spacing** > *type* **.1** > **Enter** > **Done/Return** > **Done/Return** > **Annotate** tab > [A Note] > **Make Note** > [icon] pick a point for the note in the largest area of the title block [Fig. 11.22(b)] > *type* **TOOL ENGINEERING CO.** > **Enter** > **Enter** >

**Make Note** > [icon] > *type* **DRAWN** > **Enter** > **Enter** >

**Make Note** > [icon] > *type* **ISSUED** > **Enter** > **Enter** >

**Make Note** > [icon] > *type* **&dwg_name** > **Enter** > **Enter** >

**Make Note** > [icon] > *type* **&scale** > **Enter** > **Enter** >

**Make Note** > [icon] > *type* **SHEET &current_sheet OF &total_sheets** > **Enter** > **Enter** > **Done/Return** > **LMB** to deselect > **File** > **Options** > **Sketcher** > [☐ Snap to grid] > **OK** > **No** > modify some of the text height (**.10**) and the placement of the notes as needed [Fig. 11.22(b)] > **LMB** to deselect > [icon] > [icon] > **Ctrl+S** > **Enter** [Fig. 11.22(c)] > **File** > **Close**

**Figure 11.22(b)** Parameters and Labels in the Title Block, Smaller Text is **.10** in Height

**Figure 11.22(c)** Format Notes

**Figure 11.22(d)** Title Block

**Figure 11.22(e)** Completed Drawing

Projects without instructions are available at *www.cad-resources.com.*

**Figure 12.1(a)** Clamp_Assembly Drawing

## OBJECTIVES

- Create an **Assembly Drawing**
- Generate a **Parts List** from a **Bill of Materials (BOM)**
- **Balloon** an assembly drawing
- Create a **section assembly view** and change **component visibility**
- Add **Parameters** to parts
- Create a **Table** to generate a parts list automatically
- Use **Multiple Sheets**
- Make assembly **Drawing Sheets** with **multiple models**

## REFERENCES AND RESOURCES

For **Resources** go to **www.cad-resources.com** > click on the Creo Parametric 2.0 Book cover

- Lesson 12 Lecture
- Lesson 12 3D models embedded in a PDF
- Book Projects PDF
- Project Lectures

**Figure 12.1(b)** Clamp_Subassembly Drawing

## ASSEMBLY DRAWINGS

Creo Parametric 2.0 incorporates a great deal of functionality into drawings of assemblies [Figs. 12.1(a-b)]. You can assign parameters to parts in the assembly that can be displayed on a *parts list* in an assembly drawing. Creo Parametric 2.0 can also generate the item balloons for each component on standard orthographic views or on an exploded view.

In addition, a variety of specialized capabilities allow you to alter the manner in which individual components are displayed in views and in sections. The format for an assembly is usually different from the format used for detail drawings. The most significant difference is the presence of a Parts List.

A parts list is actually a *Drawing Table object* that is formatted to represent a bill of materials in a drawing. By defining *parameters* in the parts in your assembly that agree with the specific format of the parts list, you make it possible for Creo Parametric 2.0 to add pertinent data to the assembly drawing's parts list automatically as components are added to the assembly. After the parts list and parameters have been added, Creo Parametric 2.0 can balloon the assembly drawing automatically.

In this lesson, you will create a set of assembly formats and place your standard parts list in them.

**Figure 12.2** Swing Clamp Assembly and Subassembly

## Swing Clamp Assembly Drawing

The format for an assembly drawing is usually different from the format used for detail drawings. The most significant difference is the presence of a parts list. You will create a standard "E" size format and place a standard parts list on it. You should create a set of assembly formats on "B", "C", "D", and "F" size sheets at your convenience.

A parts list is actually a *Drawing Table object* that is formatted to represent a bill of material (BOM) in a drawing. By defining parameters in the parts in your assembly that agree with the specific format of the parts list, you make it possible for Creo Parametric 2.0 to add pertinent data to the assembly drawing parts list automatically, as components are added to the assembly.

After you create an "E" size format sheet with a parts list table, you will create two new drawings (each with two views) using your new assembly format. The Swing Clamp *subassembly* [Fig. 12.2(right)] will be used in the first drawing. The second drawing will use the Swing Clamp *assembly* [Fig. 12.2(left)]. Both drawings use the "E" size format created in the first section of this lesson. The format will have a parameter-driven title block and an integral parts list.

Launch **Creo Parametric 2.0** > **Select Working Directory** > select the directory where the **clamp_assembly.asm** was saved > **OK** > **Ctrl+O** > Common Folders **System Formats** > **e.frm** > **Open** > **File** > **Save As** > **Save a Copy** > *type a unique name for your format:* New Name **asm_format_e** > **OK** > **File** > **Close** > **File** > **Open** > pick **asm_format_e.frm** > **Open**

The **format** will have a **.dtl** associated with it, and the **drawing** will have a unique **.dtl** file associated with it. *They are separate .dtl files.* When you activate a drawing and then add a format, the **.dtl** for the format controls the font, etc. for the format only. The drawing **.dtl** file that controls items on the drawing needs to be established separately in the Drawing mode.

Click: **File** > **Prepare** > **Drawing Properties** > **change** (Options dialog box opens) > *text_height* *.25* > **Enter** > *default_font* *filled* > **Enter** > *arrow_style* *filled* > **Add/Change** > *draw_arrow_length* *.25* > **Add/Change** > *draw_arrow_width* *.08* > **Add/Change** (Fig. 12.3) > **Apply** > 🖻 > File name: **asm_format_properties** > **OK** (Fig. 12.3) > **Close** > **Close** > **Ctrl+S** > **OK**

| | Value | Default | Status | Description |
|---|---|---|---|---|
| **Active Drawing** | | | | |
| arrow_style | closed * | closed | ● | Controls style of arrow head for all detail items invol |
| default_bom_balloon_type | simple_circle | | ● | Determines the default type of BOM balloon when in |
| default_font | filled.ndx | font.ndx | ● | Specifies the default text font. Include the extension |
| default_table_column_width | 10 | | ● | Specifies the default value of the column width in the |
| default_table_columns | 4 | | ● | Specifies the default number of columns of new dra |
| default_table_row_height | 1 | | ● | Specifies the default value of the row height in the n |
| default_table_rows | 2 | | ● | Specifies the default number of rows of new drawir |
| draft_scale | 1.000000 * | 1.000000 | ● | Determines value of draft dimensions relative to actu |
| draw_arrow_length | 0.250000 | 0.187500 | ● | Sets length of leader line arrows. |
| draw_arrow_width | 0.080000 | 0.062500 | ● | Sets width of leader line arrows. Drives these: "draw |
| draw_attach_sym_height | DEFAULT * | default | ● | Sets height of leader line slashes, integral signs, and |
| draw_attach_sym_width | DEFAULT * | default | ● | Sets width of leader line slashes, integral signs, and |
| draw_dot_diameter | DEFAULT * | default | ● | Sets diameter of leader line dots. If set to "default," u |
| drawing_units | inch | | ☒ | Sets units for all drawing parameters. |
| leader_elbow_length | 0.250000 * | 0.250000 | ● | Determines length of leader elbow (the horizontal leg |
| line_style_standard | std_ansi * | std_ansi | ● | Controls text color in drawings. Unless set to "std_a |
| node_radius | DEFAULT * | default | ● | Sets the size of the nodes displayed in symbols. |
| sort_method_in_region | delimited * | delimited | ● | Determines repeat regions sort mechanism. String_o |
| sym_flip_rotated_text | no * | no | ● | If set to "yes," then for new symbol definitions which |
| text_height | 0.250000 | 0.156250 | ● | Sets the default text height for newly created annota |
| text_thickness | 0.000000 * | 0.000000 | ● | Sets default text thickness for new text after regene |
| text_width_factor | 0.800000 * | 0.800000 | ● | Sets default ratio between the text width and text he |
| yes_no_parameter_display | true_false * | true_false | ● | Controls display of "yes/no" parameters in drawing r |

**Figure 12.3** New Format Options

Zoom into the title block region. Create notes for the title text and parameter text required to display the proper information.

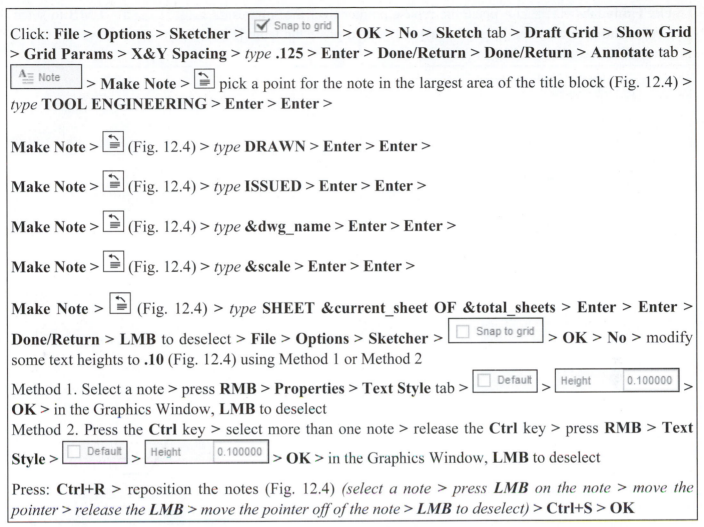

Click: **File > Options > Sketcher >** ☑ Snap to grid **> OK > No > Sketch** tab **> Draft Grid > Show Grid > Grid Params > X&Y Spacing >** *type* **.125 > Enter > Done/Return > Done/Return > Annotate** tab > A≡ Note **> Make Note >** 📄 pick a point for the note in the largest area of the title block (Fig. 12.4) > *type* **TOOL ENGINEERING > Enter > Enter >**

**Make Note >** 📄 (Fig. 12.4) > *type* **DRAWN > Enter > Enter >**

**Make Note >** 📄 (Fig. 12.4) > *type* **ISSUED > Enter > Enter >**

**Make Note >** 📄 (Fig. 12.4) > *type* **&dwg_name > Enter > Enter >**

**Make Note >** 📄 (Fig. 12.4) > *type* **&scale > Enter > Enter >**

**Make Note >** 📄 (Fig. 12.4) > *type* **SHEET &current_sheet OF &total_sheets > Enter > Enter > Done/Return > LMB** to deselect **> File > Options > Sketcher >** ☐ Snap to grid **> OK > No >** modify some text heights to **.10** (Fig. 12.4) using Method 1 or Method 2

Method 1. Select a note > press **RMB > Properties > Text Style** tab > ☐ Default **>** Height 0.100000 **> OK >** in the Graphics Window, **LMB** to deselect

Method 2. Press the **Ctrl** key > select more than one note > release the **Ctrl** key > press **RMB > Text Style >** ☐ Default **>** Height 0.100000 **> OK >** in the Graphics Window, **LMB** to deselect

Press: **Ctrl+R** > reposition the notes (Fig. 12.4) *(select a note > press LMB on the note > move the pointer > release the LMB > move the pointer off of the note > LMB to deselect)* **> Ctrl+S > OK**

**Figure 12.4** Parameters and Labels in the Title Block, Smaller Text is **.10** in Height

The parts list table can now be created and saved with this format. You can add and replace formats and still keep the table associated with the drawing. Next, start the parts list by creating a table.

Click: **Table** tab > 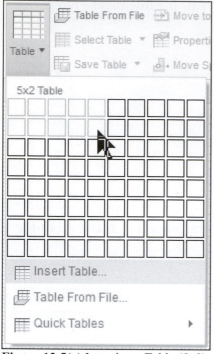 > place the mouse pointer over the squares [table grid] > **LMB** to select two rows and five columns [Fig. 12.5(a)] > **LMB** to place the table [Figs. 12.5(b-c)] > move the mouse pointer off of the table > **LMB** to deselect

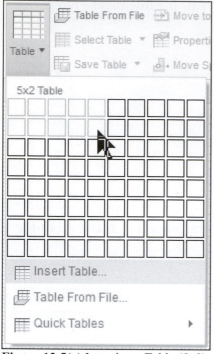

**Figure 12.5(a)** Inserting a Table (5x2)

**Figure 12.5(b)** Place the Table

**Figure 12.5(c)** Table Temporarily Placed

Position the pointer over the first column until it highlights > double-click [Fig. 12.5(d)] > change the column's width to **1.000** [Fig. 12.5(e)] > **OK**

**Figure 12.5(d)** Highlight Column

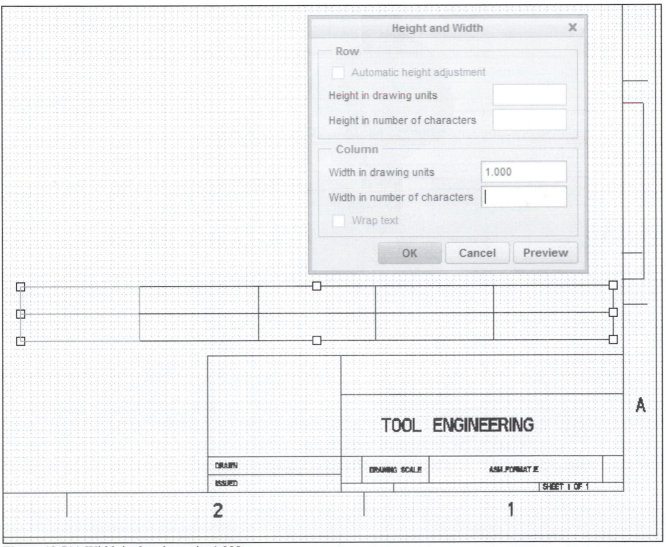

**Figure 12.5(e)** Width in drawing units **1.000**

Change the second column's width to **1.000** > change the third column's width to **4.000** [Fig. 12.5(f)] > change the fourth column's width to **1.000** > change the fifth column's width to **.750**

**Figure 12.5(f)** Width of the Third Column is **4.000**

Position the pointer over the top row (left side) until it highlights > double-click > **Automatic height adjustment** (uncheck) > change the column's height to **.375** [Fig. 12.5(g)] > **OK**

**Figure 12.5(g)** First Row Height is **.375**

Position the pointer over the top corner [Fig. 12.5(h)] > press and hold down the **LMB** > move the pointer to move the table [Fig. 12.5(i)] > release the **LMB** > move the pointer > **LMB** to deselect [Fig. 12.5(j)]

**Figure 12.5(h)** Place the Pointer on the Upper Left Corner of the Table

**Figure 12.5(i)** Move the Pointer to Position the Table

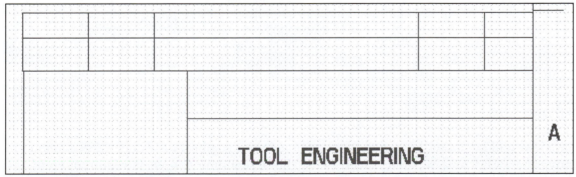

**Figure 12.5(j)** Positioned Table

Click: **Sketch** tab > **Draft Grid** > **Hide Grid** > **Done/Return** > **Table** tab > 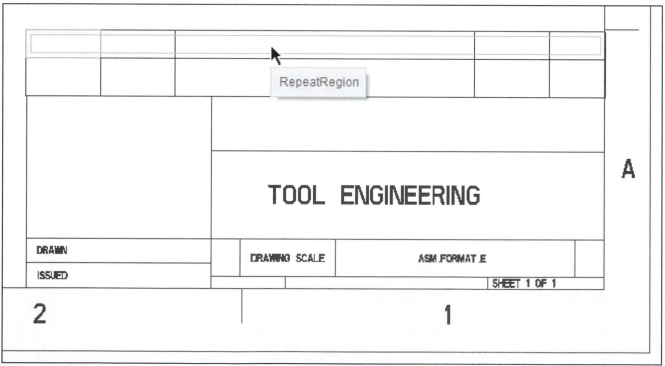 **Repeat Region** > **Add** > **Simple** > pick in the left block > pick in the right block [Fig. 12.5(k)] > **Attributes** > select the Repeat Region just created [Fig. 12.5(l)] > **No Duplicates** > **Recursive** > **Done/Return** > **Done** > **Ctrl+S** > **OK**

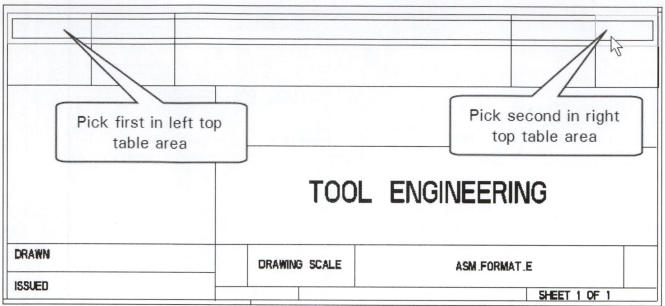

**Figure 12.5(k)** Create a Repeat Region

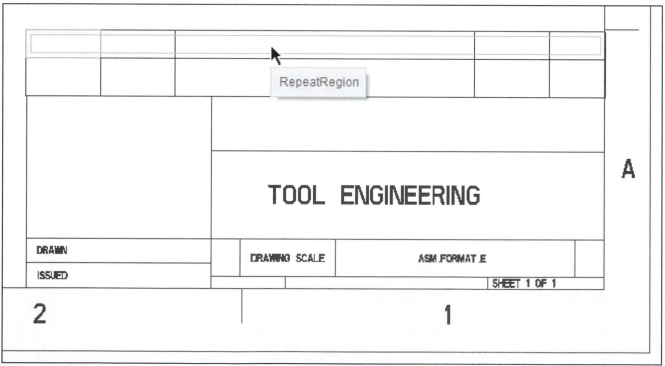

**Figure 12.5(l)** Repeat Region

Normally, a BOM will be ascending unless the drawing shows a subassembly. Also, in some cases a company will move the BOM to a separate sheet where the descending format is required. Here, we wanted you to move and resize the BOM table and become comfortable with the editing process.

Insert column headings using plain text, double-click on the first block of the second row [Fig. 12.6(a)] > *type* **ITEM** [Fig. 12.6(b)] (the text is case sensitive) > **OK**

**Figure 12.6(a)** Select the First Block

**Figure 12.6(b)** ITEM

The Repeat Region now needs to have some of its headings correspond to the parameters that will be created in each component model.

Double-click on the second block > *type* **PT NUM** [Fig. 12.6(c)] > **OK** > double-click on the third block > *type* **DESCRIPTION** > **OK** > double-click on the fourth block > *type* **MATERIAL** > **OK** > double-click on the fifth block > *type* **QTY** > **OK** [Fig. 12.6(d)]

**Figure 12.6(c)** PT NUM

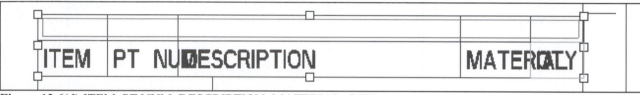

**Figure 12.6(d)** ITEM, PT NUM, DESCRIPTION, MATERIAL, QTY

With the **Ctrl** key pressed, select all five text cells > release the **Ctrl** key > press **RMB** > **Text Style** Text Style dialog box opens [Fig. 12.6(e)] > Character- Height **.125** > Note/Dimension- Horizontal **Center** > Vertical **Middle** > **Apply** [Figs. 12.6(f-g)] > **OK** > **Ctrl+S** > **OK**

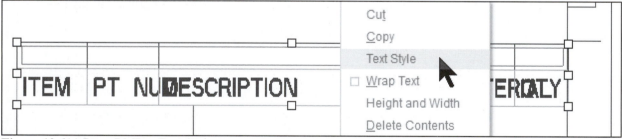

**Figure 12.6(e)** Press RMB > Text Style

**Figure 12.6(f)** Text Style Preview

**Figure 12.6(g)** Text Style Dialog Box

Insert parametric text (Report Symbols) into the table cells of the Repeat Region. Double-click on the first table cell of the Repeat Region (the table cell ABOVE the text ITEM) > **rpt...** from the Report Symbol dialog box [Fig. 12.7(a)] > **index** [Figs. 12.7(b-c)]

**Figure 12.7(a)** Report Symbol Dialog Box, Click **rpt...**

**Figure 12.7(b)** Click **index**

531

**Figure 12.7(c) &rpt.index**

Double-click on the fifth table cell of the Repeat Region > **rpt...** [Fig. 12.7(d)] > **qty** [Figs. 12.7(e-f)]

**Figure 12.7(d)** Click **rpt...**

**Figure 12.7(e)** Click **qty**

**Figure 12.7(f) &rpt.qty**

Double-click on the third (middle) table cell of the Repeat Region > **asm...** [Fig. 12.7(g)] > **mbr...** [Fig. 12.7(h)] > **User Defined** [Fig. 12.7(i)] > Enter symbol text: *type* **DSC** > **Enter** [Fig. 12.7(j)]

**Figure 12.7(g)** Click **asm...**

**Figure 12.7(h)** Click **mbr...**

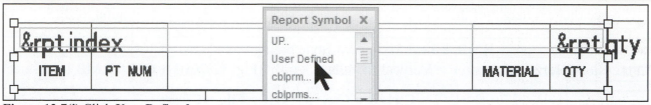

**Figure 12.7(i)** Click **User Defined**

**Figure 12.7(j) &asm.mbr.DSC**

Double-click on the fourth table cell of the Repeat Region > **asm…** > **mbr…** > **ptc_material…** > **PTC_MATERIAL_NAME** [Figs. 12.7(k-l)] > double-click on the second table cell of the Repeat Region > **asm…** > **mbr…** > **User Defined** > Enter symbol text: *type* **PRTNO** > **Enter** [Fig. 12.7(m)] > in the Graphics Window, **LMB** to deselect > **Ctrl+S** > **Enter**

**Figure 12.7(k) asm… > mbr… > ptc_material… > PTC_MATERIAL_NAME**

**Figure 12.7(l) &asm.mbr.ptc_material.PTC_MATERIAL_NAME**

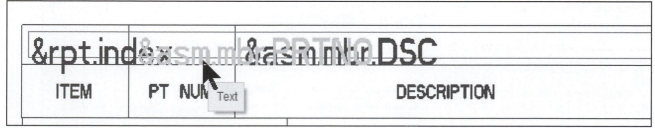

**Figure 12.7(m) &asm.mbr.PRTNO**

Change the text height of the Report Symbols in the table cells of the Repeat Region to **.125**.

Window in the table > press **RMB** > **Text Style** Text Style dialog box opens > Character- Height **.125** > Note/Dimension- Horzontal **Center** > Vertical **Middle** > **Apply** [Fig. 12.7(n)] > **OK** > **LMB** to deselect > 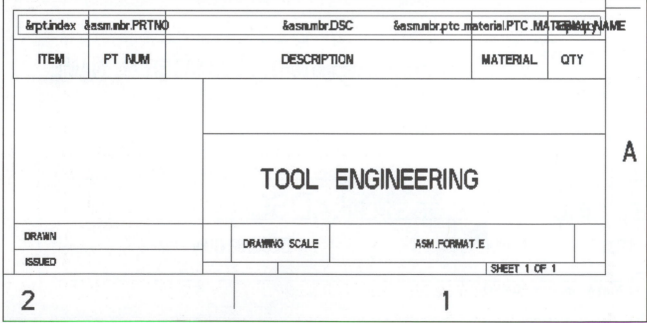 Update Tables [Fig. 12.7(o)] > > **Ctrl+R** > **Ctrl+S** > **OK** > **File** > **Close**

**Figure 12.7(n)** Text Style Dialog Box

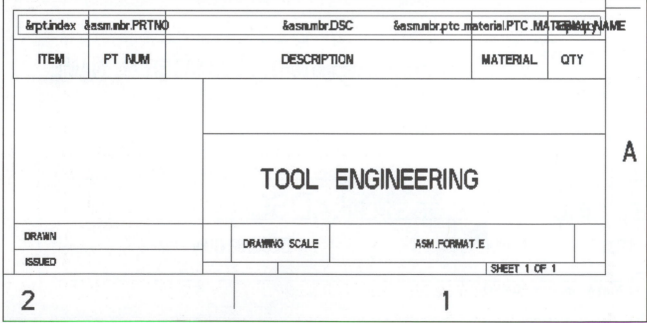

**Figure 12.7(o)** Title Block and BOM Table

534

## Adding Parts List (BOM) Data

When you save your standard assembly format, the Drawing Table that represents your standard parts list is now included. You must be aware of the titles of the parameters under which the data is stored, so that you can add them properly to your components. As you add components to an assembly, Creo Parametric 2.0 reads the parameters from them and updates the parts list. You can also see the same effect by adding these parameters after the drawing has been created. Creo Parametric 2.0 also provides the capability of displaying Item Balloons on the first view that was placed on the drawing. To improve their appearance, you can move these balloons to other views and alter the locations where they attach.

Retrieve the clamp arm, click: **File > Open > clamp_arm.prt > Open > The generic > Open > File > Options > Model Display >** Default model orientation: **Isometric > OK > No > Ctrl+D >** [icon] > [☐ (Select All)] *off >* **LMB >** [icon] *off* [Fig. 12.8(a)] > **Tools** tab > **Parameters** [Fig. 12.8(b)]

**Figure 12.8(a)** Clamp_Arm

**Figure 12.8(b)** Parameters Dialog Box

535

Click: **Parameters** from Parameters dialog box > **Add Parameter** [Fig. 12.8(c)] > Name field should be highlighted (if not, click in field and highlight) [Fig. 12.8(d)], *type* **PRTNO** [Fig. 12.8(e)] > Designate ☑ > click in Type field [Fig. 12.8(f)] > ▾ [Fig. 12.8(g)] > **String**

**Figure 12.8(c)** Add Parameter

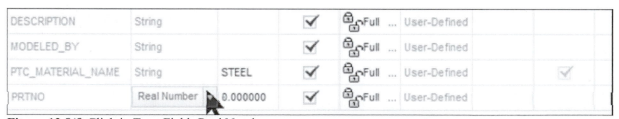

**Figure 12.8(d)** Adding Parameters

| DESCRIPTION | String | | ☑ | 🔒Full ... | User-Defined | | |
| MODELED_BY | String | | ☑ | 🔒Full ... | User-Defined | | |
| PTC_MATERIAL_NAME | String | STEEL | ☑ | 🔒Full ... | User-Defined | ☑ | |
| PRTNO | Real Num... | 0.000000 | ☑ | 🔒Full ... | User-Defined | | |

**Figure 12.8(e)** Name PRTNO

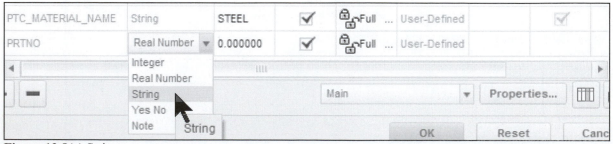

**Figure 12.8(f)** Click in Type Field- Real Number

**Figure 12.8(g)** String

536

Click in Value field > *type* **SW101-5AR** [Fig. 12.8(h)] > click in Description field > *type* **part number** [Fig. 12.8(i)] > for PTC_MATERIAL_NAME, click: Designate ☑

| DESCRIPTION | String | | ☑ | 🔒Full ... | User-Defined | | |
| MODELED_BY | String | | ☑ | 🔒Full ... | User-Defined | | |
| PTC_MATERIAL_NAME | String | STEEL | ☑ | 🔒Full ... | User-Defined | | ☑ |
| PRTNO | String | SW101-5AR | ☑ | 🔒Full ... | User-Defined | | |

**Figure 12.8(h)** Value SW101-5AR (text is case sensitive – text appears in the drawing BOM as entered here)

**Figure 12.8(i)** Add Description (describes the type of parameter – not shown in the drawing BOM)

Click: ➕ **Add new Parameter** [Fig. 12.8(j)] > in Name field, *type* **DSC** > click in Type field > ▾ > **String** > click in Value field > *type* **CLAMP ARM** > Designate ☑ > click in Description field > *type* **part description** [Fig. 12.8(k)] > **OK** > **Family Table** > 🖽 **Verify** > **Verify** > **Close** > **OK** > **Ctrl+S** > **Enter**

| PTC_MATERIAL_NAME | String | STEEL | ☑ | 🔒Full ... | User-Defined | |
| PRTNO | String | SW101-5AR | ☑ | 🔒Full ... | User-Defined | part number |
| PARAMETER_1 | Real Number | 0.000000 | ☐ | 🔒Full ... | User-Defined | |

**Figure 12.8(j)** Add New Parameter

| PTC_MATERIAL_NAME | String | STEEL | ☑ | 🔒Full ... | User-Defined | | ☑ |
| PRTNO | String | SW101-5AR | ☑ | 🔒Full ... | User-Defined | part number | |
| DSC | String | CLAMP ARM | ☑ | 🔒Full ... | User-Defined | part description | |

**Figure 12.8(k)** Add Description

Parameters can also be displayed using the Relations dialog box. In the case of the Clamp_Arm, there was a relation created for controlling a features location. Click: **Relations** > ▼ Local Parameters [Fig. 12.8(l)] > **OK > Ctrl+S > OK > File > Close**

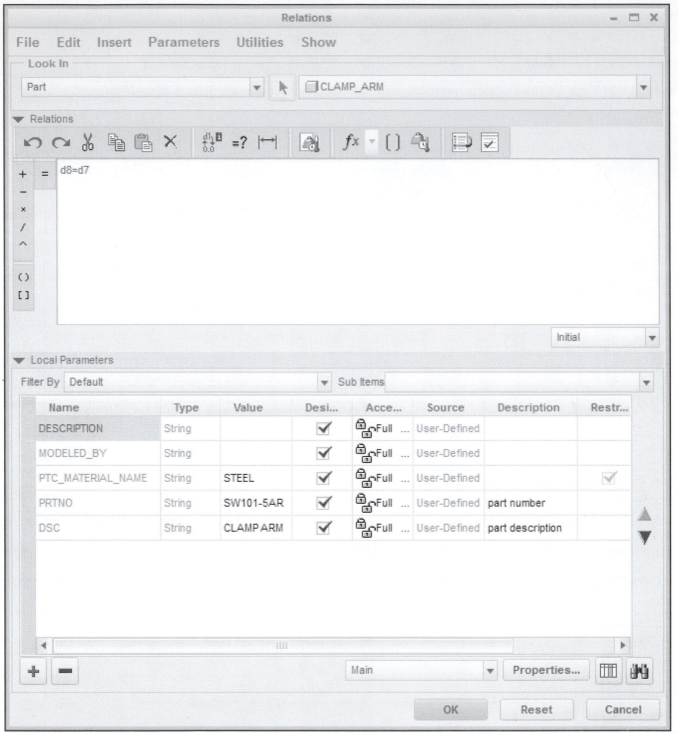

**Figure 12.8(l)** Relations Dialog Box *(your relation values may be different)*

Retrieve the clamp swivel, click: **File > Open > clamp_swivel.prt > Open > Tools** tab **> Parameters > Parameters > Add Parameter >** complete the parameters as shown (Fig. 12.9) (remember to *type* the appropriate text in the Description fields) **> OK > File > Save > Enter > File > Close**

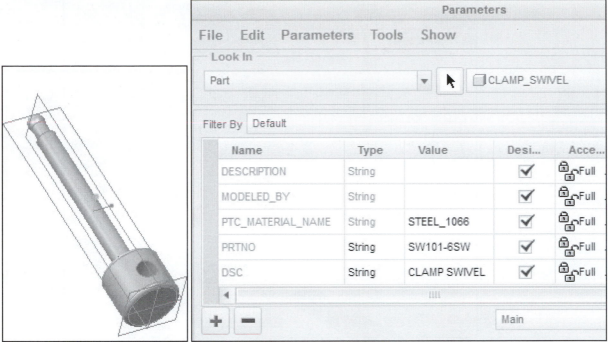

**Figure 12.9** Clamp_Swivel Parameters

Retrieve the clamp ball, click: **File > Open > clamp_ball.prt > Open > Tools** tab **> Parameters > Parameters > Add Parameter >** complete the parameters as shown (Fig. 12.10) (remember to *type* the appropriate text in the Description fields) **> OK > File > Save > Enter > File > Close**

**Figure 12.10** Clamp_Ball Parameters

Retrieve the clamp foot, click: **File > Open > clamp_foot.prt > Open > Tools** tab **> Parameters > Parameters > Add Parameter >** complete the parameters as shown (Fig. 12.11) (remember to *type* the appropriate text in the Description fields) **> OK > File > Save > Enter > File > Close**

**Figure 12.11** Clamp_Foot Parameters

Retrieve the clamp plate, click: **File > Open > clamp_plate.prt > Open > File > Prepare > Model Properties >** Material **change > steel.mtl >** ▶▶▶ [Fig. 12.12(a)] **> OK > Close > File > Save > Enter**

**Figure 12.12(a)** Materials Dialog Box

Click:  > 🔲 Tree Filters.. > toggle all Display options *on* > **OK** > **View** tab > [icons] *on* > [icons] *on* > pick **DTM3** from the model > press **RMB** > **Properties** > double-click in the Name field- *type* **A** > **Enter** > [🔲◀] [Fig. 12.12(b)] > **On Datum** > **OK** [Fig. 12.12(c)] > pick **DTM2** > press **RMB** > **Properties** > double-click in the Name field- *type* **B** > **Enter** > [🔲◀] > **On Datum** > **OK** > pick **DTM1** > press **RMB** > **Properties** > double-click in the Name field- *type* **C** > **Enter** > [🔲◀] > **On Datum** > **OK** > **LMB** to deselect

**Figure 12.12(b)** Datum Dialog Box      **Figure 12.12(c)** Set Datum Tag Annotation A

Click: **Tools** tab > **Parameters** > **Parameters** > **Add Parameter** > complete the parameters as shown [Fig. 12.12(d)] > **OK** > **File** > **Save** > **OK** > **File** > **Close**

**Figure 12.12(d)** Clamp_Plate Parameters (remember to type the appropriate text in the Description fields)

Click: **File > Open > carrlane_500_fn.prt > Open > File > Prepare > Model Properties >** Material change **> File > New** [Fig. 12.13(a)] **>** Name field- *type* **purchased** [Fig. 12.13(b)] **> Save to Library**

**Figure 12.13(a)** Materials Dialog Box

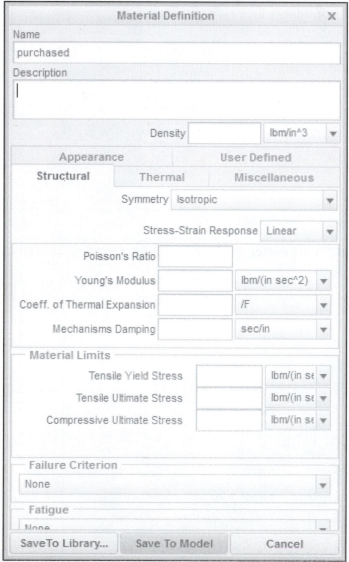

**Figure 12.13(b)** Material Definition Dialog Box

Navigate to the correct directory where you want the material to be saved, click: **OK** [Fig. 12.13(c)] > navigate to the correct directory (you should be able to click on **Working Directory**) > **purchased.mtl** > ▶▶▶ [Fig. 12.13(d)] > **OK** > **Close** > **Tools** tab > 🗔 Model [Fig. 12.13(e)] > 🖉 to close the browser

**Figure 12.13(c)** Save a Copy Dialog Box

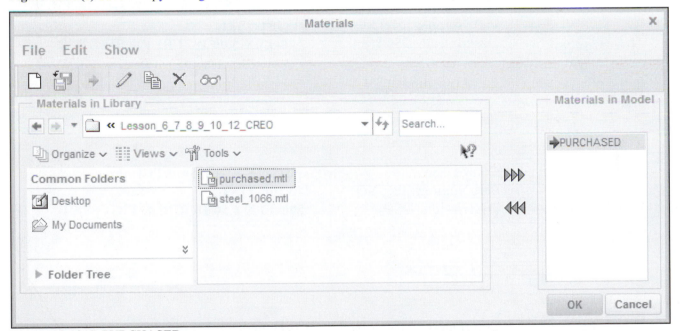

**Figure 12.13(d)** PURCHASED

| PART NAME : | CARRLANE_500_FN | | | | | |
|---|---|---|---|---|---|---|
| MATERIAL FILENAME: PURCHASED | | | | | | |
| Units: | Length: | | Mass: | Force: | Time: | Temperature: |
| Inch lbm Second (Creo Parametric Default) | in | | lbm | in lbm / sec^2 | sec | F |

**Figure 12.13(e)** Model Info, MATERIAL FILENAME: PURCHASED

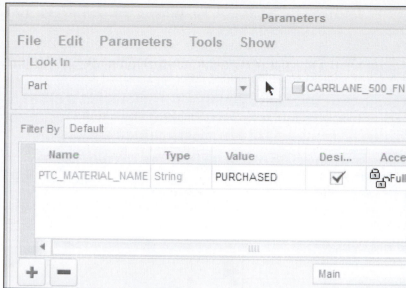

**Figure 12.13(f)** Carrlane_500_fn Parameters

Parameters can be added, deleted, and modified in Part Mode, Assembly Mode, or Drawing Mode. You can also add *parameter columns* to the Model Tree in Assembly Mode, and edit the parameter value.

| Component | Carrlane-12-13500_STUD.prt |
|---|---|
| Part Number (PRTNO) | SW101-9STL |
| Description (DSC) | .500-13 X 5.00 DOUBLE END STUD |

| Component | Carrlane-12-13350_STUD.prt |
|---|---|
| Part Number (PRTNO) | SW100-21ST |
| Description (DSC) | .500-13 X 3.50 DOUBLE END STUD |

| Component | Carrlane-500_FN.prt |
|---|---|
| Part Number (PRTNO) | SW100-22FLN |
| Description (DSC) | .500-13 HEX FLANGE NUT |

Press: **Ctrl+O** > **Carrlane-12-13350_STUD.prt** [Fig. 12.14(a)] > **Open** > **File** > **Prepare** > **Model Properties** > Material **change** > navigate to the correct directory where you saved the *purchased.mtl* material (you should be able to click on **Working Directory**) > **purchased.mtl** >  [Fig. 12.14(b)] > **OK** > **Close** > **Tools** tab > **Parameters** > Designate ☑ > **OK** > **Ctrl+S** > **Enter** > **File** > **Close**

CARRLANE-12-13350_STUD.PRT
- 350_STUD
- Neutral id 4
- Insert Here

**Figure 12.14(a)** Carrlane-12-13350_STUD.prt

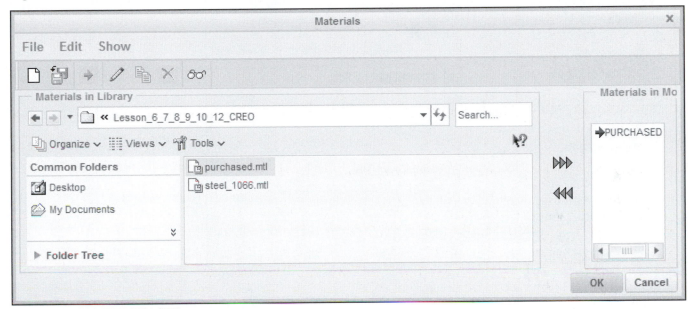

**Figure 12.14(b)** PURCHASED

Press: **Ctrl+O** > **Carrlane-12-13500_STUD.prt** [Fig. 12.15(a)] > **Open** > **File** > **Prepare** > **Model Properties** > Material **change** > navigate to the correct directory where you saved the *purchased.mtl* material (**Working Directory**) > **purchased.mtl** > ⯮ > **OK** > **Close** > **Tools** tab > **Parameters** > Designate ☑ > **OK** > 🖫 > **OK** > 🗔 Model [Fig. 12.15(b)] > 🗗 to close the browser > **File** > **Close**

CARRLANE-12-13500_STUD.PRT
- 500_STUD
- Neutral id 4
- Insert Here

**Figure 12.15(a)** Carrlane-12-13500_STUD.prt

| PART NAME : | CARRLANE-12-13500_STUD | | | | | | |
|---|---|---|---|---|---|---|---|
| MATERIAL FILENAME: PURCHASED | | | | | | | |
| Units: | Length: | | | Mass: | Force: | Time: | Temperature: |
| Inch lbm Second (Creo Parametric Default) | in | | | lbm | in lbm / sec^2 | sec | F |

**Figure 12.15(b)** Model Info

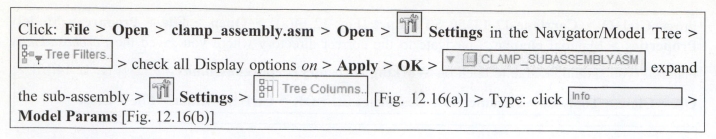

Click: **File** > **Open** > **clamp_assembly.asm** > **Open** > 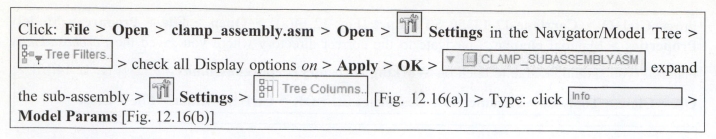 **Settings** in the Navigator/Model Tree > Tree Filters. > check all Display options *on* > **Apply** > **OK** > ▾ CLAMP_SUBASSEMBLY.ASM expand the sub-assembly > **Settings** > Tree Columns.. [Fig. 12.16(a)] > Type: click Info > **Model Params** [Fig. 12.16(b)]

**Figure 12.16(a)** Tree Columns

**Figure 12.16(b)** Model Params

Double-click in the Name field > *type* **PRTNO** [Fig. 12.16(c)] >  **Add column** > double-click in the Name field > *type* **DSC** [Fig. 12.16(d)] > **Add column** [Fig. 12.16(e)]

**Figure 12.16(c)** Type in the Name field: PRTNO

**Figure 12.16(d)** Type in the Name field: DSC

**Figure 12.16(e)** PRTNO and DSC

Double-click in the Name field > *type* **PTC_MATERIAL_NAME** [Fig. 12.16(f)] > [>>] **Add column** [Fig. 12.16(g)] > **Apply** > **OK** [Fig. 12.16(h)]

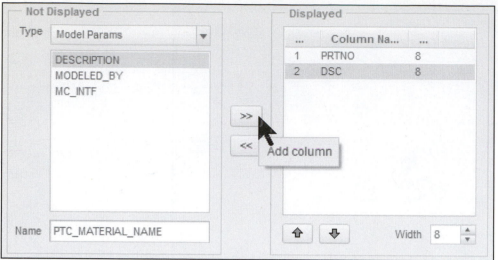

**Figure 12.16(f)** Type in the Name field: PTC_MATERIAL_NAME

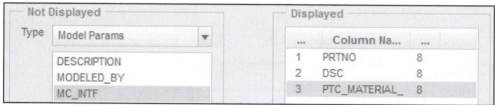

**Figure 12.16(g)** PTC_MATERIAL_NAME

| | PRTNO | DSC | PTC_MATERIAL_NAME |
|---|---|---|---|
| CLAMP_ASSEMBLY.ASM | | | |
| ▶ Annotations | | | |
| ⧅ CL_ASM_RIGHT | | | |
| ⧅ CL_ASM_TOP | | | |
| ⧅ CL_ASM_FRONT | | | |
| ⁂ CL_ASM_CSYS | | | |
| ▶ CLAMP_PLATE.PRT | SW100-20PL | CLAMP PLATE | STEEL |
| ▼ CLAMP_SUBASSEMBLY.ASM | | | |
| ▶ Placement | | | |
| ⧅ ASM_RIGHT | | | |
| ⧅ ASM_TOP | | | |
| ⧅ ASM_FRONT | | | |
| ⁂ SUB_ASM_CSYS | | | |
| ▶ CLAMP_ARM_DESIGN<CL | SW101-5AR | CLAMP ARM | STEEL |
| ▶ CLAMP_SWIVEL.PRT | SW101-6SW | CLAMP SWIVEL | STEEL_1066 |
| ▶ CLAMP_FOOT.PRT | SW101-8FT | CLAMP FOOT | STEEL |
| ▶ CARRLANE-12-13500_ST | SW101-9STL | .500-13 X 5.00 DOUBLE END STUD | PURCHASED |
| ▶ CLAMP_BALL.PRT | SW101-7BA | CLAMP BALL | NYLON |
| ▶ CLAMP_BALL.PRT | SW101-7BA | CLAMP BALL | NYLON |
| ➜ Insert Here | | | |
| ▶ CARRLANE-12-13350_STUD. | SW100-21ST | .500-13 X 3.50 DOUBLE END STUD | PURCHASED |
| ▶ CARRLANE_500_FN.PRT | SW100-22FLN | .500-13 HEX FLANGE NUT | PURCHASED |

**Figure 12.16(h)** New Tree Columns *(your tree columns may appear differently)*

Resize the Model Tree and Model Tree columns. From the Model Tree, pick on **CLAMP_SUBASSEMBLY.ASM** > click in the **PRTNO** field [Fig. 12.17(a)] > Type: click  Real Number ▾ > **String** > click in the Value field, *type* **SW101-SUB-ASM** > ☑ Designated [Fig. 12.17(b)] > click in the Description field, *type* **part number** > **OK** > click in the DSC field > Type: Real Number ▾ > **String** > click in the Value field, *type* **CLAMP SUB-ASSEMBLY** > ☑ Designated > click in the Description field, *type* **part description** > **OK** > **LMB** in the Graphics Window to deselect > **File** > **Save** > **OK** > **Ok** (if needed)

Figure 12.17(a) New Clamp Subassembly Parameter String for PRTNO

Figure 12.17(b) SW101-SUB-ASM

Complete any missing parameters for the remaining components. Where necessary modify the values to reflect what is shown here by clicking in the value field and editing the field (Fig. 12.18) (remember to click: [✓ Designated] and type the appropriate text in the Description fields) > **LMB** to deselect > double-click [⧉] to close the Navigator/Model Tree panel to display just the Graphics Window > **Ctrl+S** > **Enter** > **Ok** (if needed) > **File** > **Close**

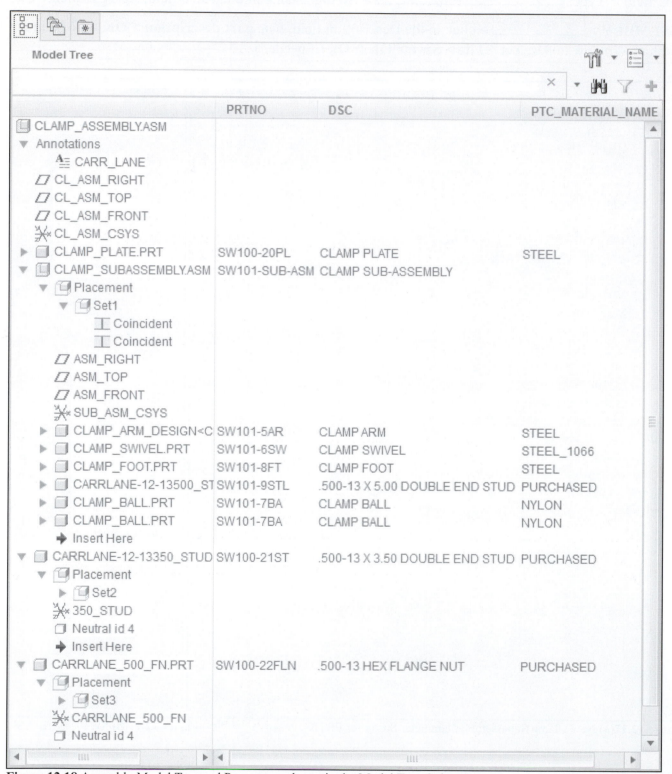

**Figure 12.18** Assembly Model Tree and Parameters shown in the Model Tree Columns

## Assembly Drawings

The parameters (and their values) have been established for each part. The assembly format with related parameters in a parts list table has been created and saved in your (format) directory. You can now create a drawing of the assembly, where the parts list will be generated automatically, and the assembly ballooned. The first assembly drawing will be of the ***Clamp_Subassembly.***

Click: **File** > **Manage Session** > **Select Working Directory** > check to see if you are in the correct working directory > **OK** > 🗋 **Create a new model** > ⦿ 🖳 Drawing > Name **clamp_subassembly** > ☐ Use default template > **OK** [Fig. 12.19(a)] > Specify Template ⦿ Empty with format > Default Model **Browse** [Fig. 12.19(b)] > **clamp_subassembly.asm** [Fig. 12.19(c)] > **Open**

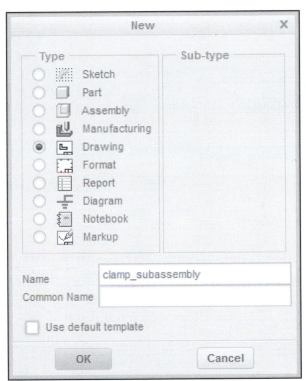

**Figure 12.19(a)** New Dialog Box

**Figure 12.19(b)** New Drawing Dialog Box

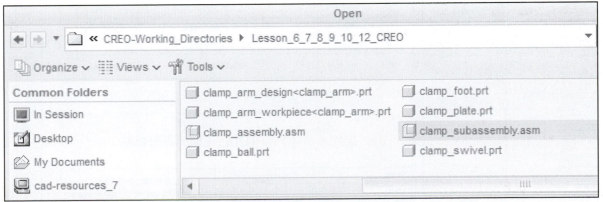

**Figure 12.19(c)** Open the clamp_subassembly.asm

551

Click: Format **Browse** [Fig. 12.19(d)] > **Working Directory** [Fig. 12.19(e)] > select **asm_format_e.frm** [Fig. 12.19(f)] > **Open** [Fig. 12.19(g)] > **OK** > **Table** tab > pan and zoom in on the title block region >  pick on a table corner to highlight > 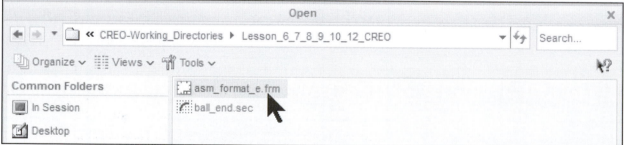 hold down the **LMB** and move the table outside of the title block [Fig. 12.19(h)] > release the **LMB** > move the pointer > **LMB** to deselect

**Figure 12.19(d)** Format Browse

**Figure 12.19(e)** Working Directory

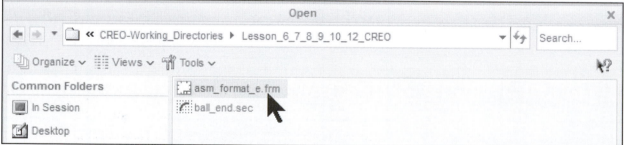

**Figure 12.19(f)** Select your previously created **asm_format_e.frm**

**Figure 12.19(g)** Format selected

| | | | | | |
|---|---|---|---|---|---|
| 1 | SW101-5AR | CLAMP ARM | STEEL | 1 |
| 2 | SW101-6SW | CLAMP SWIVEL | STEEL, ID68 | 1 |
| 3 | SW101-7BA | CLAMP BALL | NYLON | 2 |
| 4 | SW104-8FT | CLAMP FOOT | STEEL | 1 |
| 5 | SW101-9STL | .500-13 X 5.00 DOUBLE END STUD | PURCHASED | 1 |
| 6 | SW01-SUB-ASM | CLAMP SUB-ASSEMBLY | | 1 |
| ITEM | PT NUM | DESCRIPTION | MATERIAL | QTY |

TOOL ENGINEERING

A

DRAWN

ISSUED

LOGO

CLAMP SUBASSEMBLY

SHEET 1 OF 1

**Figure 12.19(h)** Title Block and BOM

Click: **File** > **Prepare** > **Drawing Properties** > Detail Options **change** > Sort: **Alphabetical** > **arrow_style filled** > **Add/Change** > crossec_arrow_length **.50** > **Enter** > crossec_arrow_width **.17** > **Enter** > default_font **filled** > **Add/Change** > draw_arrow_length **.375** > **Add/Change** > draw_arrow_width **.125** > **Enter** > max_balloon_radius **.50** > **Enter** > min_balloon_radius **.50** > **Enter** > text_height **.50** > **Add/Change** [Fig. 12.20(a)] > **Apply** >  **Save a copy of the currently displayed configuration file** > File Name **CLAMP_ASM** > **OK** [Fig. 12.20(b)] > **Close** > **Close** > **Ctrl+S** > **OK**

**Figure 12.20(a)** Drawing Options

**Figure 12.20(b)** Saving Drawing Options

If not closed, click:  to close the Navigator > [icon] > [icon] **No Hidden** > [icon] > □ (Select All) *off* > **LMB** in the Graphics Window > double-click on **SCALE: 1.000** in the lower left-hand corner of the Graphics

Window [SCALE : 1.000 TYPE : ◄◄ ◄ ✱ DrawScale] > Enter value for scale, *type* **1.50** > **Enter** > **LMB** to deselect > **Layout** tab > **Lock View Movement** unlock > in the Graphics Window, press **RMB** > **Insert General View** >

**No Combined State** (highlighted) > **OK** > [icon] > ⇨ Select CENTER POINT for drawing view. [Fig. 12.21(a)] > Model view names **FRONT** > **Apply** [Fig. 12.21(b)] > **Close** > **Ctrl+S** > **Enter**

**Figure 12.21(a)** Select Center Point for First Drawing View

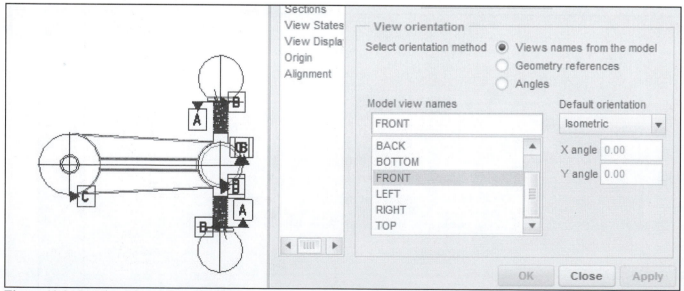

**Figure 12.21(b)** View Names from the Model

554

With the Top view still *selected/highlighted*; press: **RMB > Insert Projection View** [Fig. 12.22(a)] >
⇨ Select CENTER POINT for drawing view. (move the mouse pointer below the view previously created) [Fig.
12.22(b)] > **LMB** to select the position [Fig. 12.22(c)] > **Ctrl+S** > **OK**

**Figure 12.22(a)** Insert Projection View

**Figure 12.22(b)** Location below the Previous View

**Figure 12.22(c)** Selected position

With the Front view still *selected/highlighted*, press: **RMB** > **Properties** [Fig. 12.23(a)] > **Sections** > 
[ ⦿ 2D cross-section ] > [ + ] [Fig. 12.23(b)] > [ ✓ A ] > **OK** [Fig. 12.23(c)] > **LMB** > **Ctrl+S** > **OK**

**Figure 12.23(a)** Properties

**Figure 12.23(b)** Drawing View 2D section (Section A should have been previously created)

**Figure 12.23(c)** SECTION A_A

Click:  open the Navigator > **Annotate** tab > from the Drawing Tree expand Datums ▼ ✕ᴬ Datums > with the **Ctrl** key (or **Shift** key) pressed, select all of the Model datums > release the **Ctrl** (or **Shift**) key > **RMB** > **Erase** > repeat for the other view [Fig. 12.24(a)] > **LMB** in the Graphics Window to deselect > reposition the views as needed [Fig. 12.24(b)] > 🔍 > **Ctrl+S** > **OK** > pick on the Front (lower) view to select/highlight > **Layout** tab > in the Graphics Window, press **RMB** > **Add Arrows** [Fig. 12.25(a)] > pick on the Top (upper) view > move the cursor outside of the view outline > **LMB** [Fig. 12.25(b)]

**Figure 12.24(a)** RMB > Erase    **Figure 12.24(b)** Reposition Views

**Figure 12.25(a)** Press RMB > Add Arrows

**Figure 12.25(b)** Section Arrows

Creo Parametric 2.0 provides tools to alter the display of the section views to comply with industry ASME standard practices. Most companies require that the crosshatching on parts in section views of assemblies be "clocked" such that parts that meet do not use the same section lining (crosshatching) spacing and angle. This makes the separation between parts more distinct. First, modify the visibility of the views to remove hidden lines and make the tangent edges dimmed.

---

Press and hold down the **Ctrl** key > pick on both views > release the **Ctrl** key > with the cursor outside of the view outlines, press **RMB** > **Properties** > Display style [No Hidden] > Tangent edges display style [Dimmed] > Hidden line removal for xhatches [● Yes] [Fig. 12.26(a)] > **Apply** > **Close** > **LMB** in the Graphics Window to deselect [Fig. 12.26(b)]

---

**Figure 12.26(a)** No Hidden

**Figure 12.26(b)** View Display

Click: **Annotate** tab >  > tab > All > Axes [Fig. 12.26(c)] > with the **Ctrl** key pressed, pick both views > release the **Ctrl** key > in the Show column, next to each required axis; click ✓ *(as you move the mouse pointer over each entry in the list, it highlights in the Graphics Window)* > **OK** > **LMB** to deselect [Fig. 12.26(d)] > 🔍 > 💾 > **OK**

**Figure 12.26(c)** Type Axes

**Figure 12.26(d)** Displaying Axes

Click: **Layout** tab > double-click on the crosshatching in the Front view (CARRLANE-12-13500_STUD is now active) > **Fill** [Figs. 12.26(e-f)] > click **Next** until CLAMP_FOOT is active > **Delete Line** > **Hatch** > **Spacing** > **Double** > **Angle** [Fig. 12.26(g)] > **135** [Fig. 12.26(h)]

**Figure 12.26(e)** Modify Xhatch    **Figure 12.26(f)** CARRLANE-12-13500_STUD Fill Section

**Figure 12.26(g)** Angle **135**    **Figure 12.26(h)** CLAMP_FOOT Hatching

Click: **Next** until CLAMP_SWIVEL is active > **Angle** > **45** > **Next** until CLAMP_ARM is active > **Delete Line** > **Angle** > **120** [Fig. 12.26(i)] > **Previous** until CLAMP_SWIVEL is active > **Spacing** > **Half** > **Done** > in the Graphics Window, **LMB** to deselect > 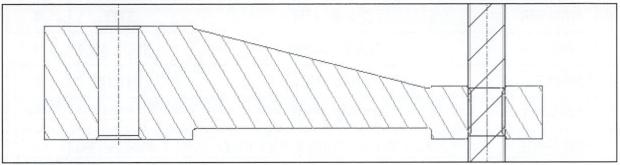 [Fig. 12.26(j)] > **OK**

**Figure 12.26(i)** CLAMP_ARM, Hatch Spacing Angle **120**

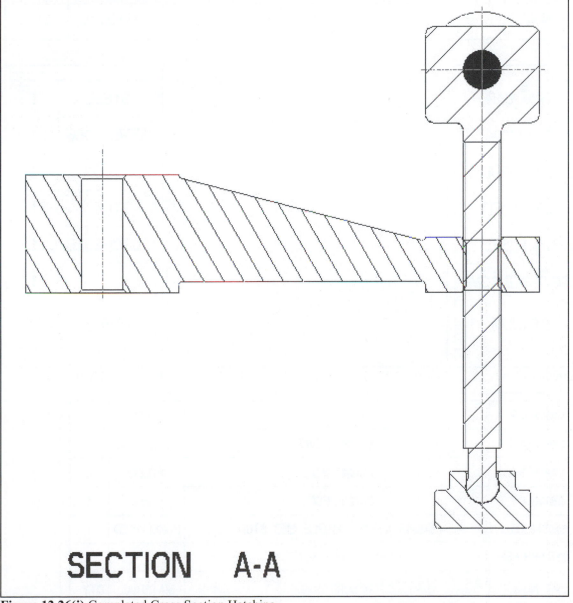

SECTION   A-A

**Figure 12.26(j)** Completed Cross Section Hatching

Zoom in on the BOM [Fig. 12.27(a)] > **Table** tab > place the pointer to highlight the PT NUM column [Fig. 12.27(b)] > double-click > Width in drawing units **1.25** [Fig. 12.27(c)] > **OK** > place the pointer to highlight the QTY column [Fig. 12.27(d)] > double-click > Width in drawing units **.50** > OK [Fig. 12.27(e)] > **LMB** to deselect

| 1 | SW101-5AR | CLAMP ARM | STEEL | 1 |
|---|---|---|---|---|
| 2 | SW101-6SW | CLAMP SWIVEL | STEEL .1066 | 1 |
| 3 | SW101-7BA | CLAMP BALL | NYLON | 2 |
| 4 | SW101-8FT | CLAMP FOOT | STEEL | 1 |
| 5 | SW101-9STL | .500-13 X 5.00 DOUBLE END STUD | PURCHASED | 1 |
| 6 | SW101-SUB-ASM | CLAMP SUB-ASSEMBLY | | 1 |
| ITEM | PT NUM | DESCRIPTION | MATERIAL | QTY |

**Figure 12.27(a)** BOM

| 1 | SW101-5AR |
|---|---|
| 2 | SW101-6SW |
| 3 | SW101-7BA |
| 4 | SW101-8FT |
| 5 | SW101-9STL |
| 6 | SW101-SUB-ASM |
| ITEM | PT NUM |

**Figure 12.27(b)** Highlight Column

**Height and Width**

Row
- [ ] Automatic height adjustment
- Height in drawing units [        ]
- Height in number of characters [        ]

Column
- Width in drawing units [ 1.250 ]
- Width in number of characters [ 5.3 ]
- [ ] Wrap text

OK    Cancel    Preview

**Figure 12.27(c)** Height and Width Dialog Box

| STEEL | 1 |
|---|---|
| STEEL .1066 | 1 |
| NYLON | 2 |
| STEEL | 1 |
| PURCHASED | 1 |
| | 1 |
| MATERIAL | QTY |

**Figure 12.27(d)** Qty Column

| 1 | SW101-5AR | CLAMP ARM | STEEL | 1 |
|---|---|---|---|---|
| 2 | SW101-6SW | CLAMP SWIVEL | STEEL .1066 | 1 |
| 3 | SW101-7BA | CLAMP BALL | NYLON | 2 |
| 4 | SW101-8FT | CLAMP FOOT | STEEL | 1 |
| 5 | SW101-9STL | .500-13 X 5.00 DOUBLE END STUD | PURCHASED | 1 |
| 6 | SW101-SUB-ASM | CLAMP SUB-ASSEMBLY | | 1 |
| ITEM | PT NUM | DESCRIPTION | MATERIAL | QTY |

**Figure 12.27(e)** Adjusted BOM Table

To complete the drawing, the balloons must be displayed for each component. Balloons are displayed in the top view as the default, because it was the first view that was created.

Click: [image] > [-5] **Create Balloons > Create Balloons - All** [Fig. 12.28(a)] > press and hold the **Ctrl** key and pick on the balloons for the Clamp_Arm (1), Clamp_Swivel (2) and Clamp_Foot (4) > release the **Ctrl** key > press **RMB** > **Move Item to View** [Fig. 12.28(b)]

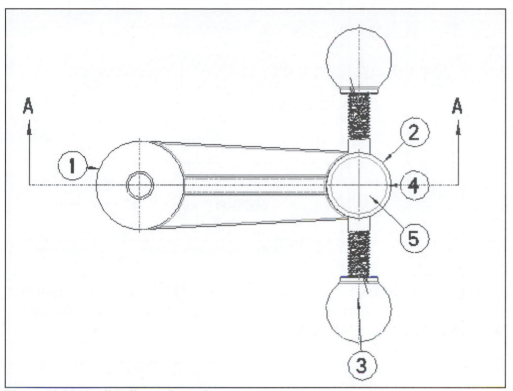

**Figure 12.28(a)** All Balloons Created

**Figure 12.28(b)** Move Item to View

Pick on the Front (lower) view [Fig. 12.28(c)] > move the pointer outside of the view boundary > **LMB** to deselect > pick on and reposition each balloon as needed > pick on balloon **1** > press **RMB** > **Edit Attachment** > **On Entity** > pick a new position > **MMB** > reposition the balloon [Fig. 12.28(d)] > in the Graphics Window, **LMB** to deselect > **Ctrl+S** > **MMB**

Figure 12.28(c) Balloons Moved to Front View

Figure 12.28(d) Repositioned Balloons in the Front View

Pick on balloon **5** (CARRLANE-12-13500_STUD) > press **RMB** [Fig. 12.28(e)] > **Edit Attachment** > **On Entity** > pick a new position on the CARRLANE-12-13500_STUD [Fig. 12.28(f)] > **Done/Return** > reposition the balloon > in the Graphics Window, **LMB** to deselect

Figure 12.28(e) Edit Attachment

Figure 12.28(f) Select the Edge of the STUD

564

Reposition and edit the attachment of all ballons as necessary >  [Figs. 12.29(a-b)] > 🖫 > **OK** > **File > Manage File > Delete Old Versions >** ✓ **> File > Close > File > Manage Session > Erase Not Displayed > OK**

SECTION A-A

**Figure 12.29(a)** Completed Clamp_Subassembly Drawing

| 1 | SW101-5AR | CLAMP ARM | STEEL | 1 |
| 2 | SW101-6SW | CLAMP SWIVEL | STEEL .1066 | 1 |
| 3 | SW101-7BA | CLAMP BALL | NYLON | 2 |
| 4 | SW101-8FT | CLAMP FOOT | STEEL | 1 |
| 5 | SW101-9STL | .500-13 X 5.00 DOUBLE END STUD | PURCHASED | 1 |
| 6 | SW101-SUB-ASM | CLAMP SUB-ASSEMBLY | | 1 |
| ITEM | PT NUM | DESCRIPTION | MATERIAL | QTY |

**Figure 12.29(b)** Clamp_Subassembly BOM

The Swing Clamp Assembly drawing is composed of the subassembly, the plate, the short stud, and the nut. The drawing will use the same format created for the subassembly. Formats are normally read-only files that can be used as many times as needed.

Click: **Ctrl+N** > ◉ 🖳 Drawing > Name **CLAMP_ASSEMBLY** > ☐ Use default template > **OK** > Default Model **Browse** > **clamp_assembly.asm** [Fig. 12.30(a)] > **Open** > ◉ Empty with format > Format **Browse** > Common Folders **Working Directory** > select **asm_format_e.frm** [Fig. 12.30(b)] > **Open** [Fig. 12.30(c)] > **OK**

**Figure 12.30(a)** Open clamp_assembly.asm

**Figure 12.30(b)** asm_format_e.frm

**Figure 12.30(c)** New Drawing Dialog Box

Zoom in on the title block > **Table** tab > window-in to select the table > move table above the title block > change the PT NUM column width to **1.25** and the QTY column width to **.50** as per the sub-assembly drawing > [Fig. 12.31(a)] > 🔍 > 🖊 > **File** > **Prepare** > **Drawing Properties** > Detail Options **change** > 📂 **Open a configuration file** > clamp_asm.dtl > **Open** > **Apply** > **Close** > **Close** > 💾 > **Enter**

| ITEM | PT NUM | DESCRIPTION | MATERIAL | QTY |
|------|--------|-------------|----------|-----|
| 1 | SW100-20PL | CLAMP PLATE | STEEL | 1 |
| 2 | SW100-21ST | .500-13 X 3.50 DOUBLE END STUD | PURCHASED | 1 |
| 3 | SW100-22FLN | .500-13 HEX FLANGE NUT | PURCHASED | 1 |
| 4 | SW101-5AR | CLAMP ARM | STEEL | 1 |
| 5 | SW101-6SW | CLAMP SWIVEL | STEEL .1066 | 1 |
| 6 | SW101-7BA | CLAMP BALL | NYLON | 2 |
| 7 | SW101-8FT | CLAMP FOOT | STEEL | 1 |
| 8 | SW101-9STL | .500-13 X 5.00 DOUBLE END STUD | PURCHASED | 1 |
| 9 | SW101-SUB-ASM | CLAMP SUB-ASSEMBLY | | 1 |

TOOL ENGINEERING

A

| DRAWN | | | |
|-------|---------|----------------|-------------|
| ISSUED | 1.000 | CLAMP_ASSEMBLY | |
| | | | SHEET 1 OF 1 |

2                                          1

**Figure 12.31(a)** BOM and Title Block

Double-click on **SCALE: 1.000** in the lower left-hand corner of the Graphics Window `SCALE : 1.000` >

Enter value for scale

`1.50` > ✓ > **LMB** to deselect > 🔍 **Refit**

Click: **Layout** tab > in the Graphics Window, press **RMB** > **Insert General View** > **No Combined State** (highlighted) [Fig. 12.31(b)] > **OK** >  (pick where you want a Top view) > **Ctrl+R** > Model view names **FRONT** [Fig. 12.31(c)] > **Apply** [Fig. 12.31(d)] > **Close** > in the Graphics Window, press **RMB** > **Lock View Movement** *off* (if already unchecked > **RMB** in the Graphics Window)

**Figure 12.31(b)** No Combined State

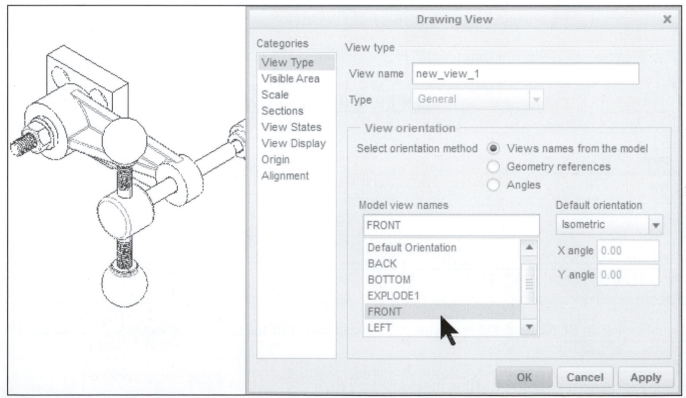

**Figure 12.31(c)** Drawing View Dialog Box

**Figure 12.31(d)** Top View Placed

Press: **RMB > Insert Projection View** [Fig. 12.32(a)] > pick a position for the Front view

**Figure 12.32(a)** Press RMB > Insert Projection View

Press: **RMB** > **Properties** [Fig. 12.32(b)] > Categories **Sections** > ⦿ 2D cross-section > ➕
( Create New... ▾ ) [Fig. 12.32(c)] > **Planar** > **Single** > **Done** > Enter NAME for cross-section, *type* **A** >
**Enter**

**Figure 12.32(b)** View Properties

**Figure 12.32(c)** Add Cross-Section to View

Click:  > ☑ 🗐 Plane Display > **LMB** in the Graphics Window > **View** tab > 🖾 *on* > 🖾 > **Layout**

tab > ⇨ Select or create an assembly datum pick on **CL_ASM_TOP** *(from the Top view)* [Fig. 12.32(d)] >

**Apply** > 🗐 > □ 🗐 Plane Display > **LMB** in the Graphics Window > **View** tab > 🖾 🖾 🖾 🖾 *off* >

🔍 [Fig. 12.32(e)] > **Close** > in the Graphics Window, **LMB** to deselect > **Annotate** tab > if not open,

click: 🗐 to open the Navigator > from the Navigator Model Tree, click: 🗐 > 🗐 Tree Filters... > toggle

all Display options *on* > **OK**

**Figure 12.32(d)** Select the CL_ASM_TOP Datum Plane

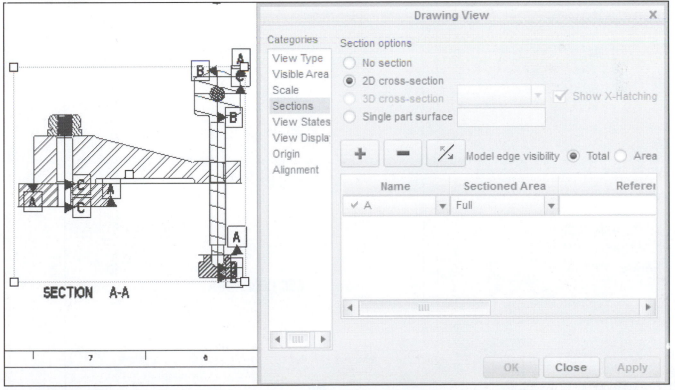

**Figure 12.32(e)** Categories: Sections

From the Drawing Tree, click: [ ▼ ⟩⟨ Datums ] (to expand) > with the **Ctrl** key (or **Shift** key) pressed, select all of the Model datums > release the **Ctrl** (or **Shift**) key > **RMB** > **Erase** > repeat for lower view [Fig. 12.32(f)] > in the Graphics Window, **LMB** to deselect

**Figure 12.32(f)** Drawing Tree Model Datums > RMB > Erase

Pick on the Front (lower) view > press **RMB** > **Add Arrows** [Fig. 12.32(g)] >
⇨ Pick a view for arrows where the section is perp. MIDDLE button for none. pick on the Top (upper) view > move the
cursor outside of the view outline > **LMB** [Fig. 12.32(h)] > **View** tab > **Ctrl+S** > **OK** > **Ok** (if needed)

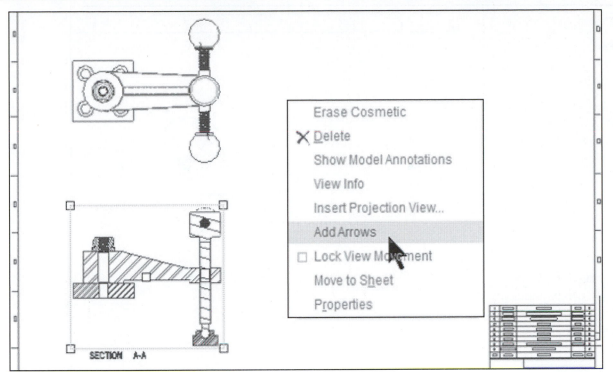

**Figure 12.32(g)** Press RMB > Add Arrows

**Figure 12.32(h)** Cutting Plane Arrows Displayed

With the **Ctrl** key pressed; in the Model Tree, select the three components [Fig. 12.32(i)] > release the **Ctrl** key > press **RMB** > **Show Model Annotations** >  >  > **OK** >  > **LMB** in the Graphics Window to deselect > **Ctrl+S** > **OK**

**Figure 12.32(i)** Assembly Drawing Views with Three Components Selected

Zoom in on the BOM, click: **Table** tab > **Repeat Region** > **Attributes** > pick in the BOM RepeatRegion field > **Flat** [Fig. 12.33(a)] > **Done/Return** > **Done** > select and move the BOM as needed [Fig. 12.33(b)]

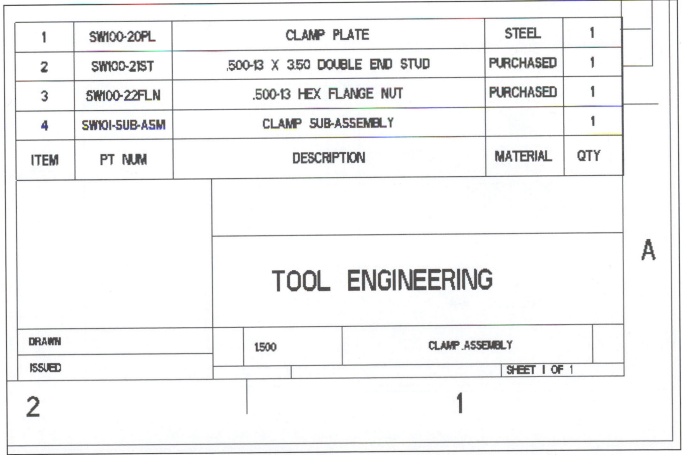

| 1 | SW100-20PL | CLAMP PLATE |
| 2 | SW100-21ST | .500-13 X 3.50 DOUBLE END STUD |
| 3 | SW100-22FLN | .500-13 HEX FLANGE NUT |
| 4 | SW101-5AR | CLAMP ARM |
| 5 | SW101-65W | CLAMP SWIVEL |
| 6 | SW101-7BA | CLAMP BALL |
| 7 | SW101-8FT | CLAMP FOOT |
| 8 | SW101-9STL | .500-13 X 5.00 DOUBLE END STUD |
| 9 | SW101-SUB-ASM | CLAMP SUB-ASSEMBLY |
| ITEM | PT NUM | DESCRIPTION |

RepeatRegion

Menu Manager
- No Duplicates
- No Dup/Level
- Recursive
- Flat
- Min Repeats
- Start Index
- No Start Idx
- Bln By Part
- Bln By Comp
- Cable Info
- No Cbl Info
- Done/Return

**Figure 12.33(a)** Table Showing with BOM Attribute Recursive, Select Flat from the Menu Manager

| 1 | SW100-20PL | CLAMP PLATE | STEEL | 1 |
| 2 | SW100-21ST | .500-13 X 3.50 DOUBLE END STUD | PURCHASED | 1 |
| 3 | SW100-22FLN | .500-13 HEX FLANGE NUT | PURCHASED | 1 |
| 4 | SW101-SUB-ASM | CLAMP SUB-ASSEMBLY | | 1 |
| ITEM | PT NUM | DESCRIPTION | MATERIAL | QTY |

A

TOOL ENGINEERING

| DRAWN | | | |
| ISSUED | 1500 | CLAMP ASSEMBLY | SHEET 1 OF 1 |

2                                    1

**Figure 12.33(b)** Repositioned BOM Table with Attribute Flat

**Figure 12.33(c)** Balloons Displayed in Top View

While pressing the **Ctrl** key, pick on balloons 1, 2, and 3 > release the **Ctrl** key > press **RMB** > **Move Item to View** > pick on the Front view [Fig. 12.33(d)] > pick on balloon **2** > press **RMB** > **Edit Attachment** > **On Entity** > pick on an edge [Fig. 12.33(e)] > **MMB** > reposition the balloon > pick on and reposition (reattach) each balloon as needed > in the Graphics Window, **LMB** to deselect

**Figure 12.33(d)** Press RMB > Move Item to View

**Figure 12.33(e)** Repositioned Balloons

Most companies (and as per drafting standards) require that rounded purchased items, such as nuts, bolts, studs, springs, and die pins be excluded from sectioning even when the section cutting plane passes through them. *If your system does not display the short stud, highlight in the Model Tree. If it does display, then exclude the section lining from it on the drawing.*

---

Click: **Layout** tab > double-click on the crosshatching in the Front view [Fig. 12.33(f)] (the CARRLANE_500_FN is now active) > **Exclude** to not show the Section of CARRLANE_500_FN > **Next** (CARRLANE-12-13350_STUD is now active) > **Exclude** > **Next** (CARRLANE-12-13500_STUD is now active) > **Fill** the cross section of the long stud > **Next** (Clamp_Foot is now active) > **Spacing** > **Double** > **Delete Line** > **Next** (Clamp_Swivel is now active) > **Angle** > **135** > **Spacing** > **Half** > **Next** (Clamp_Arm is now active) > **Delete Line** > **Spacing** > **Half** > **Next** (Clamp_Plate is now active) >

**Delete Line** > **Angle** > **120** > **Done** > in the Graphics Window, **LMB** to deselect > [🔍] > **Ctrl+S** > **OK**

---

**Figure 12.33(f)** Sectioned View

578

## Exploded Swing Clamp Assembly Drawings

The process required to place an exploded view on a drawing is similar to adding assembly orthographic views. The BOM will display all components on this sheet.

Click: **Layout** tab > New Sheet > 🔍 > from the Ribbon, **Lock View Movement** *off* > from the

Ribbon, General > **COMB0001** > OK > ➡️ Select CENTER POINT for drawing view.
> Categories- **View Type** > Model view names EXPLODE1 [Fig. 12.34(a)] > Default orientation
Isometric > **Apply** > Categories- **View Display** > Display style Shading > **OK** > 💾 > **OK**

**Figure 12.34(a)** View Type

579

With the view still *selected/highlighted*, press: **RMB > Show Model Annotations >** 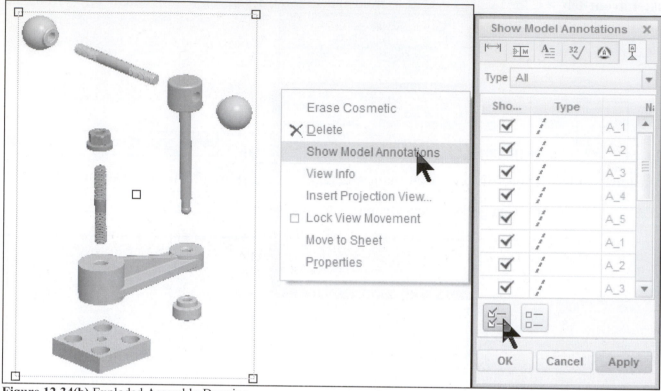 **> All >** [Fig. 12.34(b)] > **Apply > Cancel > LMB > Annotate** tab > erase extra centerlines and clip each centerline (axis) to extend between components that are in line [Fig. 12.34(c)] >  > **OK**

**Figure 12.34(b)** Exploded Assembly Drawing

A_1(AXIS):F5(REVOLVE_1):CLAMP_SWIVEL

**Figure 12.34(c)** Clip the Centerlines Axes

Click: **Table** tab >  **Create Balloons > Create Balloons - All >** ⇨ Select a region. pick in the BOM field > reposition the balloons and their attachment points as needed to clean up the drawing [Fig. 12.35(a)] > pick on pick balloon **6** > press **RMB > Edit Attachment > On Surface > Dot** > pick on the Clamp_Ball's surface > **Done/Return > LMB** > zoom in on the BOM, move table above the Title Block, and adjust the columns as necessary > 🔍 > Sheet 1 > 💾 > **OK > File > Save As** > Type ▾ > **Zip File (*.zip) > OK** > upload to your course interface or attach to an email and send to your instructor

**Figure 12.35(a)** Exploded View

Combine the sub-assembly drawing with the assembly drawing, click: **Layout** tab > [Insert ▼] **Insert** > [Import Drawing/Data] > [clamp_subassembly.drw] > **Open** > [Sheet 1 | Sheet 2 | Sheet 3] > [Sheet 1 | Sheet 2 | Sheet 3] [Fig. 12.35(b)] > [Sheet 1] > [💾] > **OK** > **File** > **Close** > **File** > **Exit** > **Yes**

SCALE : 1.200   TYPE : ASSEM   NAME : CLAMP_ASSEMBLY   SIZE : E   SHEET 2 OF 3

**Figure 12.35(b)** Completed Drawing, Sheet 2

Create a *documentation package* containing all models and drawings required to manufacture the parts and assemble the components. Your instructor may change the requirements, but in general, create and plot/print the following:

- **Part Models** for all components
- **Detail Drawings** for each nonstandard component
- **Assembly Drawings** using standard orthographic ballooned views
- **Exploded Subassembly Drawing** of the ballooned subassembly
- **Exploded Assembly Drawing** of the ballooned assembly

A different assembly Project (Coupling) is available at ***www.cad-resources.com***

*In order to save space and reduce the purchase price of this book; Lessons 13 through 18 are available, at no additional cost, for download from the publisher to those purchasing this book.*

# Index

Oriented 157, 403
Origin 27, 35, 118, 147
Overlapping Geometry 279

**P**
Palette 34, 161, 185, 249
Pan 86, 355
Parallel 348
Parameter Symbols 277
Parameters 10, 92, 215, 219, 275-279
Parametric 1-4
Parametric Design 1-4
Parametric Dimensions 2
Parametric Modeling 2-4
Parent-Child Relationships 2, 8, 13,
    151, 321, 342
Parent Feature 2, 8, 149, 151
Parents 6, 8
Part Design 5-6
Part Drawings 437, 481
Part Library 75
Part Mode 16, 211
Part Number 536
Partial View 447, 497
Parts List 519, 523, 535, 551
Paste 281
Paste Special 275, 282
Pattern(s) 11, 277, 378
Pattern (Dimension) 379
Pattern (Direction) 378
Perpendicular Constraint 158, 202
Perspective View 84, 403, 416, 447
Place View 445, 476
Placement 167, 180, 336
Placement Dashboard 320
Placement Tab 24, 38-40, 125, 237,
    252, 299, 336
Placing Components 320
Planar 6, 212, 276, 362, 570
Plane Tag Display 337
Plotting 177, 274
Point 202
PopUp Viewer 93
Portrait 441
Preview 28, 36, 56, 115, 227, 355,
    369, 380, 385, 482
Primary Datum Planes 6
Print 221
Print Active Object 274
Pro/Library 67, 74, 322-329
Pro/MANUFACTURING 320
Pro/NC 320
Productivity Enhancements 100, 104
Profile Rib 276, 290, 293, 304
Projection View 446, 487, 555, 569
Properties 1, 153, 178, 197, 331, 522,
    541
Properties (Datum) 199
Properties (Layer) 197, 210
Properties (Units) 178

Protrusions 152
PTC_MATERIAL_NAME 533, 548
ptc-ceramic 124
ptc-metallic-gold 41, 47
ptc-metallic-brushed-steel 184
ptc-std-aluminum 80

**Q**
**qty** 529-532, 562, 567
Quantity 385
Quick Access Toolbar 70
Quick Fix 278
Quick Reference Cards 17, 21, 111
Quick Sash 533

**R**
Record Mapkey 107
Rectangle 113-115, 173, 367
Recursive 528, 575
Redefine 213
Reference Dimension(s) 15, 446, 495,
    508
Reference Viewer 373, 377
Referenced Features 1-4
References 37, 124, 136, 157, 172,
    189, 290, 377
References Dialog Box 156, 189
References Filters Tab 373, 377
References Tab 423
Refit 29, 36, 53, 95, 192, 462, 567
Refit (Drawing) 462, 470, 567
Regenerate 144, 163, 271, 283, 342,
    361, 503
Regenerate the Section 163
Regeneration Manager 343
Regenerating Models 342
Relation(s) 12, 14, 214, 218, 276, 309,
    317, 338
Relations and Parameters 218, 317
Remove All 375
Remove Material 128, 138, 169, 171,
    187, 295
Rename 81, 105, 108, 178, 224
Reorder 2
Repaint 19, 85, 198
Repeat Region (Table) 528-534, 575
Report Symbols 531, 534
Resize 157, 162, 346
Resolve 271, 278, 342
Resolve Feature 278
Resume 271, 278, 311
Resume All 211
Resume Layers 211
Resuming Features 209, 311, 407
Revolve 134, 137, 221, 236, 263
Revolve Tool 221-222
Revolved Cut 139, 222, 228, 242, 244
Revolved Extrusion 228, 236, 264
Revolved Feature(s) 221-222, 244,
    251

Ribbon 17, 19, 22, 24, 71, 84, 89, 123,
    157, 175, 179, 409, 442, 575
Rib Tool 275, 290, 293
Ribs 275-276
Right (view) 482, 487, 503
Right Datum Plane 24, 38, 135, 175,
    188, 263
**RMB** 18
Round(s) 1, 112, 115, 243, 276
Round Edges 116, 134, 143, 258
Round Tool 112, 116, 134, 143, 258
**rpt** 531

**S**
Sash 248, 326
Save 19, 29, 36, 131
Save a Copy 81, 241, 317, 485, 522
Save Display Elements 420
Save the Active Object 29
Saved View 416, 445
Saved View List 417
Scale 437, 445, 447, 478, 498, 514
Scale (Drawing) 445, 447, 474
Scale View 447, 498
Scheme 5, 88, 160, 184, 190
Screen Tip 71, 410
Section(s) 135, 177, 182, 210-213,
    222, 249, 363, 469, 481, 491, 495,
    501, 556
Section Lines 509, 560, 578
Section Orientation 290
Section Sketch 11, 143, 241
Section Trajectory 146
Section View 447, 469, 495, 503, 558
Sections (Models) 362
Select Instance 78, 450
Select Working Directory 22, 74, 113,
    153, 322, 482, 522
Separate Window 320, 335, 345, 350
Set Active 362, 420
Set Datum 177, 199, 255, 541
Set Layers 197
Set Region 575
Set Horizontal Reference 290
Set Hyperlink Screen Tip 410
Set Working Directory 226, 449, 482
Settings 43, 88, 179, 308, 546, 559
Shade 113, 120, 288
Shading 85, 126, 153, 188, 296, 359
Shading with Edges 80, 126
Shading with reflections 54
Shape Tab 207, 252, 265, 304, 369
Shell 4, 112, 117, 122
Sheet Setup 57, 62, 451
**Shift+MMB** 157
Shortcut Keys 18
Show 3D Dragger 345, 350, 385
Show Annotations 316
Show Component 320, 345, 350
Show Datums 113